# INFERENTIAL STATISTICS

Dr. Vandana Bagla

**Publisher:** CreateSpace Independent Publishing Platform (A unit of Amazon)
**Price:** $22.50

**Marketed By**

# *Preface*

It is my pleasure to present the first edition of the book titled 'Inferential Statistics'. The book provides an insight into elementary inferential statistical methodologies including probability and sampling techniques. The content elaborates statistical methods that are significantly useful in science and engineering undergraduate courses; engrossing basic concepts and applications of statistical techniques in scientific, research and industrial applications. With this first edition, I have tried to meet the expectations of the students by explaining the text and related theory right from the elementary level. With a substantial emphasis on conceptual knowledge, the book provides working methodologies with sufficient number of illustrative examples. Practicing students and budding scientists will find it pragmatic in research related application problems. Level of the book has been kept moderately rudimentary and plain sailed to provide its' readers with lucidity and perceptibility. An exuberant attempt has been made for the students to comprehend and develop an envision for the concepts exemplified. It is hoped that this book would be determinedly advantageous to the tutees and prove to be serviceable.

**Dr. Vandana Bagla**
B.Sc. (Computers), M.Sc. (Mathematics), MBA (HR), Ph.D. (Operations Research)
Assistant Professor at Maharaja Agrasen Institute of Technology,
Guru Gobind Singh Indraprastha University, Delhi, India

# DEDICATED TO MY LOVING FATHER
# MR. JAGAN NATH DHINGRA

# Contents

**Chapter 1: Introduction to Inferential Statistics** .................................................................................. 7

1.1 Inferential Statistics ........................................................................................................................ 7

    1.1.1 Measures of central tendencies .............................................................................................. 7

        1.1.1.1 Mean ................................................................................................................................. 7

        1.1.1.2 Median .............................................................................................................................. 7

        1.1.1.3 Mode ................................................................................................................................. 8

    1.1.2 Measures of Variation (Dispersion) ......................................................................................... 8

        1.1.2.1 Range ................................................................................................................................ 8

        1.1.2.2 Quartile Deviation ............................................................................................................. 9

        1.1.2.3 Standard Deviation ........................................................................................................... 9

        1.1.2.4 Variance ............................................................................................................................ 9

**Chapter 2: Theory of Probability** ........................................................................................................ 10

2.1 Introduction .................................................................................................................................. 10

2.2 Mathematical Definition of Probability ........................................................................................ 10

2.3 Addition Law of Probability .......................................................................................................... 14

2.4 Conditional Probability ................................................................................................................. 16

    2.4.1 Multiplicative Law of Probability ........................................................................................... 16

2.5 Bay's Theorem ............................................................................................................................... 18

2.6 Random Variable and Probability Distributions ........................................................................... 21

    2.6.1 Discrete Probability Distributions .......................................................................................... 21

    2.6.2 Continuous Probability Distributions ..................................................................................... 24

2.7 Moments ....................................................................................................................................... 26

2.8 Moment Generating Function (MGF) ........................................................................... **Error! Bookmark not defined.**

2.9 Skewness ...................................................................................................................... **Error! Bookmark not defined.**

2.10 Kurtosis ....................................................................................................................... **Error! Bookmark not defined.**

**Chapter 3: Theoretical Probability Distributions** ............................................................................... 32

3.1 Introduction .................................................................................................................................. 32

3.2 Binomial Distribution .................................................................................................................... 32

    3.2.1 Mean of Binomial Distribution ............................................................................................... 32

    3.2.2 Variance of Binomial Distribution .......................................................................................... 32

    3.2.3 Moment Generating Function of Binomial Distribution ....................................................... 33

3.3 Poisson Distribution ...................................................................................................................... 38

3.3.1 Mean of Poisson Distribution ........................................................................................... 39

3.3.2 Variance of Poisson Distribution ..................................................................................... 39

3.3.3 Moments of Poisson Distribution ................................................................................... 39

3.4 Normal Distribution ................................................................................................................. 42

3.4.1 Moments of Normal Distribution .................................................................................... 44

**Chapter 4: Curve Fitting & Correlation** ........................................................................................ 50

4.1 Introduction ............................................................................................................................. 50

4.2 Principle of Least Squares ........................................................................................................ 50

4.2.1 Fitting a straight line ........................................................................................................ 51

4.2.2 Fitting a parabola ............................................................................................................. 53

4.2.3 Change of Scale ................................................................................................................ 55

4.3 Correlation ............................................................................................................................... 57

4.4 Karl Pearson Coefficient of Correlation ................................................................................... 57

4.4.1 Shortcut Method for Karl Pearson Coefficient of Correlation ......................................... 58

4.4.2 Coefficient of Correlation of Bivariate Frequency Distribution ...................................... 59

4.4.3 Coefficient of Correlation by Rank differences ................................................................ 60

4.5.1 Properties of Regression Coefficients .............................................................................. 65

4.5.2 Angle between the Lines of Regression .......................................................................... 66

**Chapter 5: Sampling Distributions & Hypothesis Testing** ............................................................ 74

5.1 Introduction ............................................................................................................................. 74

5.2 Sampling Distributions ............................................................................................................. 76

5.3 Sampling of Attributes for large samples ($n > 30$) ............................................................... 77

5.3.1 Comparing Proportions of Large Samples from Two Different Populations in Terms of Attributes ............. 79

5.4 Sampling Distributions of Sample means ................................................................................ 80

**Chapter 6: Tests of Significance for Small Samples** ..................................................................... 89

6.1 Introduction ............................................................................................................................. 89

6.2 Student's $t$-Distribution .......................................................................................................... 89

6.3 Snedecor's $F$ – test for Testing Equality of Two Population Variances .................................. 95

6.4 Fisher's $Z$ Test for Testing Significance of Correlation Coefficient for Small Samples ........... 97

**Chapter 7: Tests of significance based on Chi-Square** ................................................................. 100

7.1 Chi-Square ($\chi 2$) Test .............................................................................................................. 100

7.2 Chi-Square ($\chi 2$) Test for Testing Significance of Sample Variance ....................................... 100

7.3 Chi-Square ($\chi 2$) Test for Testing Goodness of Fit ................................................................. 101

7.4 Chi-Square ($\chi^2$) Test for Testing Independence of Attributes ............................................................. 103

**Chapter 8: Analysis of Variance** ................................................................................................................ 108

8.1 Introduction .............................................................................................................................................. 108

8.2 One-Way ANOVA ...................................................................................................................................... 108

8.3 Two-Way ANOVA ...................................................................................................................................... 115

**Appendix** ......................................................................................................................................................... 121

Scilab code to find mean, S.D. and first $r$ moments about mean of a grouped data ............................... 121

Scilab code to find mean, S.D. and first $r$ moments about mean of an ungrouped grouped data ........ 121

Scilab code to find a line of regression ...................................................................................................... 123

Scilab code to fit a straight line to given set of data points .................................................................... 123

Scilab code to fit a parabola to given set of data points ......................................................................... 124

Table1: Area Under Standard Normal Curve ............................................................................................. 125

Table2: Standard Normal Cumulative Probability Table .......................................................................... 126

Table3: *t*-table ............................................................................................................................................. 127

Table4: Chi-Square ($\chi^2$) Distribution Table ............................................................................................ 128

Table 5: Critical Values of the F Distribution with Alpha Level of 0.05 .................................................. 129

*Chapter 1: Introduction to Inferential Statistics*

# Introduction to Inferential Statistics

## 1.1 Inferential Statistics

Statistics is an important branch of mathematics dealing with data collection, analysis and interpretation. Statistics has vast scientific and research applications and can be divided into two parts: descriptive statistics and inferential statistics. Both of these are employed in data analysis and are of equal importance in practical applications.

By definition an inference is an educated guess or conclusion that is drawn from evidences and reasoning. Inferential statistics as the name suggests, involves drawing the conclusions from the statistical analysis that has been performed using descriptive statistics. Inferences are made about central tendency and variability or other aspects of a distribution and this is generally done through random sampling.

### 1.1.1 Measures of central tendencies

Statisticians often use common parameters to describe a set of observations which measure how the set is centered around a particular point on a line scale. This category of parameters is called measures of central tendencies and commonly used measures are mean, median and mode. The most reliable statistical parameter from this category is the mean or average.

#### 1.1.1.1 Mean

The mean is the arithmetic average of a set of observations. The mean gives an idea where the center lies for a set of observations. The arithmetic mean is obtained by taking the sum of all the observations in the set and dividing by the total number of observations in the set.

Mean of a sample is said to be an unbiased estimator of the population mean and is highly useful in more advanced statistical interpretations drawn in inferential statistics. It is an inevitable parameter of inferential statistics.

**For ungrouped data:** Mean $(\bar{x}) = \frac{\sum_{i=1}^{n} x_i}{n} = \frac{x_1 + x_2 + \cdots + x_n}{n}$

**For grouped data:** Mean $(\bar{x}) = \frac{\sum f_i x_i}{\sum f_i}$, $\sum f_i = n$ **(In case of frequency distribution)**

Mean $(\bar{x}) = a + \frac{\sum f_i d_i}{\sum f_i} \times h$, $\sum f_i = n$ **(In case of class intervals)**

here $a$ is the assumed mean and $h$ is height of the class interval and $d_i = \frac{x_i - a}{h}$.

#### 1.1.1.2 Median

Although mean is the most widely used measure of central tendency, it is not always appropriate to use it; especially when the set of observations has few one sided extreme (highest or lowest) values. Mean is strongly influenced by the few peripheral observations in skewed distributions. In such cases mean may not truthfully represent the central tendency of the set of observations because it has been raised (or lowered) by a few outlying scores. In such situations median may be a better measure of central tendency. The median value in a set of observations is that value which divides the set into equal halves when all the numbers have been arranged in ascending or descending order.

**For ungrouped data:** After arranging the data in ascending or descending order,

$$\text{Median} = \begin{cases} \text{middle term,} & \text{when } n \text{ is odd} \\ \text{mean of two middle terms,} & \text{when } n \text{ is even} \end{cases}$$

**For grouped data:** Median class is determined by class in which $\frac{n}{2}$ lies

$$\text{Median} = l + \frac{\frac{n}{2} - c}{f} \times h$$

where $l$ is the lower limit of the median class, $c$ is the cumulative frequency before the median class, $f$ is the frequency of the median class, $h$ is height of the class interval.

### 1.1.1.3 Mode

The mode is another measure of central tendency and is the most frequently occurring term in the set of observations. Mode of a set of observations may not be unique.

**For ungrouped data:** Mode of a raw data is the term which has highest frequency.

**For ungrouped data:** Modal class is the class having highest frequency

$$\text{Mode} = l + \frac{f_m - f_{m-1}}{(f_m - f_{m-1}) + (f_m - f_{m+1})} \times h$$

where $l$ is the lower limit of the modal class, $f_m$ is the frequency of the modal class, $f_{m-1}$ is the frequency of the class before the modal class, $f_{m+1}$ is the frequency of the class after the modal class, $h$ is height of the class interval.

## 1.1.2 Measures of Variation (Dispersion)

Apart from measuring central tendencies to summarise a set of data, we need to measure the degree of variation (dispersion) of the set to determine its reliability. Measures of central tendency are useful in understanding how the observations cluster about a central value and measures of dispersion are useful in understanding how far the extreme (high or low) observations are scattered about the central value of the set.
Common measures of variation (dispersion) are: range, quartile deviation, standard deviation and variance.

### 1.1.2.1 Range

The range is the simplest of the measures of variation and describes the difference between the lowest value and the highest value in a set of observations.

Range = (largest value of the set) − (smallest value of the set)

For grouped data, the range is the difference between the upper bound of the highest class interval and the lower bound of the lowest class interval.

Range is very simple to use but it provides relatively little information on dispersion as it is highly influenced by an extremely large or an extremely small observation in the data set.

### 1.1.2.2 Quartile Deviation

If $Q_1, Q_2, Q_3$ are the quartiles which divide the data (arranged in ascending or descending order) into four equal parts, then Inter quartile range $= Q_3 - Q_1$.

Moreover

$Q_1$ is the median of the lower half of the data
$Q_2$ is the median of the whole set of data
$Q_3$ is the median of the upper half of the data

Quartile Deviation $(Q.D.)$ is usually taken as measure of dispersion for the data whose central tendency is measured as median and is given by $Q.D. = \frac{1}{2}(Q_3 - Q_1)$.

### 1.1.2.3 Standard Deviation

Along with the mean, the standard deviation is an inevitable part of inferential statistics. Quartile deviation is used in association with the median whereas standard deviation is measured for data whose central tendency is taken as mean. The standard deviation gives an approximate picture of the average amount each number in a set varies from the central value (mean).

Standard deviation $(\sigma)$ is defined as root mean square deviation, i.e. $\sigma = \sqrt{\frac{\Sigma(x_i - \bar{x})}{n}}$

**In case of grouped data**

In case of frequency distribution

$$\sigma = \sqrt{\frac{\Sigma f_i (x_i - \bar{x})}{\Sigma f_i}}, \quad \Sigma f_i = n$$

In case of class intervals

$$\sigma = \sqrt{\frac{\Sigma f_i d_i^2}{\Sigma f_i} - \left(\frac{\Sigma f_i d_i}{\Sigma f_i}\right)^2} \times h, \quad d_i = \frac{x_i - a}{h}, \quad h \text{ is height of the class interval}$$

### 1.1.2.4 Variance

Variance is the square of standard deviation. It is denoted by $\sigma^2$

*Chapter 2: Theory of Probability*

# Theory of Probability

## 2.1 Introduction

The numerical measure of certainty of an event is called probability. The probability of any event lies between 0 and 1. Probability of a sure event is 1 while that of an impossible event is 0.

**Sample Space**: The set of all possible outcomes associated with an experiment is called sample space. For example while tossing a coin, the sample space is $S = \{H, T\}$ and while tossing two coins $S = \{HH, HT, TH, TT\}$, whereas while rolling a die $S = \{1,2,3,4,5,6\}$.

**Event**: An event is the subset of a sample space. For example getting an odd number while rolling a die is an event $E = \{1,3,5\}$.

**Mutually Exclusive Events**: Two or more events $E_1, E_2, \ldots \ldots, E_n$ in a sample space are mutually exclusive if they have no point in common i.e. if $E_1 \cap E_2 \cap \ldots \cap E_n = \phi$. Getting an odd number and getting an even number while rolling a die are two mutually exclusive events.

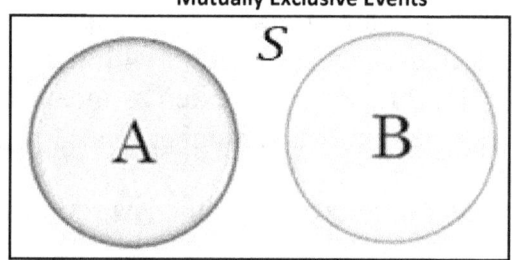

**Exhaustive Events**: Two or more events are called exhaustive events if at least one of them occurs when an experiment is performed. For example while tossing a coin; there are two exhaustive cases head and tail. Also while rolling a die; getting an odd number and getting an even number are two exhaustive events. Whereas getting a number less than two and getting a number greater than two, in case of rolling a die are non-exhaustive events.

**Mutually Exclusive and Exhaustive Events**:

If events $E_1, E_2, \ldots, E_n$ in a sample space are mutually exclusive and exhaustive then

$$P(E_1) + P(E_2) + \ldots + P(E_n) = 1$$

**Equally Likely Events**: Two or more events are equally likely if they have same probability of occurrence. Getting an odd number and getting an even number while rolling an unbiased die are two mutually exclusive and equally likely events.

**Independent Events**: Two events are said to be independent if the occurrence or non-occurrence of one does not affect the probability of occurrence of the other. Mathematically the events $E_1$ and $E_2$ are independent if and only if $P(E_1 \cap E_2) = P(E_1) \times P(E_2)$.

If two events are independent; they cannot be mutually exclusive and vice-versa.

## 2.2 Mathematical Definition of Probability

If a trial results in $n$ exhaustive, mutually exclusive and equally likely events and $m$ of them are favorable to happening of an event $E$; then probability of happening of $E$ is given by:

$$P(E) = \frac{\text{Favorable number of cases}}{\text{Exaustive number of cases}} = \frac{n(E)}{n(S)} = \frac{m}{n}$$

**Example1** Give an example of two events which are mutually exclusive but not exhaustive.

**Solution**: In an experiment of tossing two coins,

Let $E_1$: Getting two heads

$E_2$: Getting two tails

Then $S = \{HH, HT, TH, TT\}$

$E_1 = \{HH\}$ and $E_2 = \{TT\}$

$\therefore E_1 \cap E_2 = \phi$ but $P(E_1) + P(E_2) \neq 1$

Hence the events $E_1$ and $E_2$ are mutually exclusive but not exhaustive.

**Example2** Give an example of two events in an experiment of tossing two coins, which are mutually exclusive and exhaustive.

**Solution:** In an experiment of tossing two coins,

Let $E_1$: Getting at least one head

$E_2$: Getting two tails

Then $S = \{HH, HT, TH, TT\}$

$E_1 = \{HH, HT, TH\}$ and $E_2 = \{TT\}$

$\therefore E_1 \cap E_2 = \phi$ and $P(E_1) + P(E_2) = 1$

Hence the events $E_1$ and $E_2$ are mutually exclusive and exhaustive.

**Example 3** An urn contains 5 white, 6 red and 4 black balls. Two balls are drawn at random. Find the probability that both are red. Also find the probability of one white and one black ball.

**Solution**: White Balls: 5, Red Balls: 6, Black Balls: 4

Let $E$: Both balls are red

$P(E) \equiv P(\text{red ball})$ and $P(\text{red ball})$

$\Rightarrow P(E) = \frac{6}{15} \times \frac{5}{14} = \frac{1}{7}$

Let $F$: One white and one black ball are drawn

$P(F) \equiv (P(\text{white ball}) \text{ and } P(\text{black ball})) \text{ or } (P(\text{black ball}) \text{ and } P(\text{white ball}))$

$\Rightarrow P(F) = \frac{5}{15} \times \frac{4}{14} + \frac{4}{15} \times \frac{5}{14} = \frac{4}{21}$

Note: Questions in which replacement is not allowed can be attempted in a better manner using combinations.

Using combinations, $P(E) = \frac{6_{C_2}}{15_{C_2}} = \frac{6 \times 5}{15 \times 14} = \frac{1}{7}$

$P(F) = \frac{5_{C_1} \times 4_{C_1}}{15_{C_2}} = \frac{5 \times 4 \times 2!}{15 \times 14} = \frac{4}{21}$

**Example4** An urn contains 4 white, 5 red and 6 black balls. Three balls are drawn at random. Find the probability that balls are white, red and black.

**Solution**: Given that: white balls are 4, red balls are 5 and black balls are 6

Let $E$: Balls are white, red and black

$P(E) = \frac{4_{C_1} \times 5_{C_1} \times 6_{C_1}}{15_{C_3}} = \frac{4 \times 5 \times 6 \times 3!}{15 \times 14 \times 13} = \frac{24}{91}$

**Example 5** Four cards are drawn without replacement from a well shuffled pack of 52 cards. Find the probability that
  (i)   All cards are spades
  (ii)  There are two spades and two hearts
  (iii) All cards are black

Also compute the probabilities if four cards are drawn with replacement.

**Solution**: Let $E_1$: All cards are spades
$E_2$: There are two spades and two hearts
$E_3$: All cards are black

$$P(E_1) = \frac{13_{C_4}}{52_{C_4}} = \frac{13 \times 12 \times 11 \times 10 \times 4!}{52 \times 51 \times 50 \times 49 \times 4!} = \frac{11}{4165}$$

$$P(E_2) = \frac{13_{C_2} \times 13_{C_2}}{52_{C_4}} = \frac{13 \times 12 \times 13 \times 12 \times 4!}{52 \times 51 \times 50 \times 49 \times 2! \times 2!} = \frac{468}{20825}$$

$$P(E_3) = \frac{26_{C_4}}{52_{C_4}} = \frac{26 \times 25 \times 24 \times 23 \times 4!}{52 \times 51 \times 50 \times 49 \times 4!} = \frac{46}{833}$$

Again if the four cards are drawn with replacement

$$P(E_1) = \frac{13}{52} \times \frac{13}{52} \times \frac{13}{52} \times \frac{13}{52} = \frac{1}{256}$$

$$P(E_2) = \frac{13}{52} \times \frac{13}{52} \times \frac{13}{52} \times \frac{13}{52} \times \frac{4!}{2! \times 2!} = \frac{3}{128}$$

(Here $\frac{4!}{2! \times 2!}$ denotes the number of arrangements (permutations) of spades and hearts, which are SSHH or HHSS or SHSH or SHHS or HSHS or HSSH where S denotes Spade and H denotes Heart)

$$P(E_3) = \frac{26}{52} \times \frac{26}{52} \times \frac{26}{52} \times \frac{26}{52} = \frac{1}{16}$$

**Example 6** A pair of dice is rolled. What is the probability of sum 7?

**Solution**: Let E: Getting a sum 7, when a pair of dice is rolled

$$S = \begin{Bmatrix} (1,1), (1,2), \ldots, (1,6), (2,1), (2,2), \ldots, (2,6), (3,1), (3,2), \ldots, (3,6), \\ (4,1), (4,2), \ldots, (4,6), (5,1), (5,2), \ldots, (5,6), (6,1), (6,2), \ldots, (6,6) \end{Bmatrix}$$

$E = \{(1,6), (2,5), (3,4), (4,3), (5,2), (6,1)\}$

$$P(E) = \frac{n(E)}{n(S)} = \frac{6}{36} = \frac{1}{6}$$

**Example 7** Only 3 events A, B, C can happen. Given that chance of A is one-third that of B and odds against C are 2:1, find odds in favor of A.

**Solution**: Given $P(A) + P(B) + P(C) = 1$ ...①

Also $P(A) = \frac{1}{3} P(B) \Rightarrow P(B) = 3\,P(A)$ ...②

And $P(C) = \frac{1}{3}$ ...③

Using ②, ③ in ①

$\Rightarrow P(A) + 3P(A) + \frac{1}{3} = 1$

$\Rightarrow 4P(A) = \frac{2}{3}$

$\Rightarrow P(A) = \frac{1}{6}$ and $P(\bar{A}) = \frac{5}{6}$

Hence odds in favour of A are 1:5

**Example 8** A problem in mathematics is given to three students A, B and C whose chances of solving are $\frac{2}{3}, \frac{1}{2}$ and $\frac{1}{3}$ respectively. What is the probability that the problem will be solved?

**Solution**: Probability of A solving the problem $P(A) = \frac{2}{3} \Rightarrow P(\bar{A}) = \frac{1}{3}$

Probability of B solving the problem $P(B) = \frac{1}{2} \Rightarrow P(\bar{B}) = \frac{1}{2}$

Also probability of C solving the problem $P(C) = \frac{1}{3} \Rightarrow P(\bar{C}) = \frac{2}{3}$

Now probability that A, B and C do not solve the problem is $\frac{1}{3} \times \frac{1}{2} \times \frac{2}{3} = \frac{1}{9}$

∴ The probability that the problem is solved = 1 − (Probability that problem is not solved)

$$= 1 - \frac{1}{9} = \frac{8}{9}$$

**Example 9** A bag contains 50 tickets numbered from 1 to 50, out of which 5 are drawn at random and arranged in ascending order ($t_1 < t_2 < t_3 < t_4 < t_5$). Find the probability of $t_4$ carrying the number 45.

**Solution**: Exhaustive number of cases = $50_{C_5}$.

To follow the given pattern, three tickets $t_1, t_2, t_3$ must be drawn out of tickets numbered from 1 to 44 with favorable number of cases $^{44}C_3$, then $t_4$ is drawn bearing number 25 which has only one favourable case $^1C_1$, and then $t_5$ is drawn out of remaining 5 tickets with favourable number of cases $^5C_1$.

∴ Required probability is $\frac{^{44}C_3 \times ^1C_1 \times ^5C_1}{50_{C_5}} = \frac{44 \times 43 \times 42 \times 5 \times 5!}{50 \times 49 \times 48 \times 47 \times 46 \times 3!} = 0.03$

**Example 10** A has two shares in lottery in which there is 2 prizes and 3 blanks; B has three shares in a lottery in which there are 3 prizes and 6 blanks. Compare the probability of A's success to that of B's success.

**Solution**: Probability of A not getting a prize in two shares $= \frac{3_{C_2}}{5_{C_2}} = \frac{3 \times 2}{5 \times 4} = \frac{3}{10}$

∴ Probability of A getting a prize $= \frac{7}{10}$

Probability of B not getting a prize in three shares $= \frac{6_{C_3}}{9_{C_3}} = \frac{6 \times 5 \times 4}{9 \times 8 \times 7} = \frac{5}{21}$

∴ Probability of B getting a prize $= \frac{16}{21}$

Hence the probability of A's success to that of B's success $= \frac{7}{10} : \frac{16}{21}$

$$= 147 : 160$$

**Example 11** What is the probability that a leap year selected at random will have 53 Mondays.

**Solution**: A leap year has 366 days, having 52 full weeks and 2 extra days.

∴ S = {(Monday, Tuesday), (Tuesday, Wednesday), (Wednesday, Thursday),
  (Thursday, Friday), (Friday, Saturday), (Saturday, Sunday), (Sunday, Monday)}

Let E: The leap year selected will have 53 Mondays.

$E = \{(\text{Monday, Tuesday}), (\text{Sunday, Monday})\}$

$\therefore P(E) = \frac{n(E)}{n(S)} = \frac{2}{7}$

**Example 12** A and B alternatively throw a die until one gets a success and wins the game, where success is defined as getting a six. What are their respective probabilities of winning if A takes the first trial?

**Solution**: If success (S) is getting a six, then $P(S) = \frac{1}{6}$, and if F denotes failure, then $P(F) = \frac{5}{6}$.

Now A can win in $1^{st}, 3^{rd}, 5^{th}, \cdots$ trials, i.e. getting $S$ or $FFS$ or $FFFFS, \cdots$

$\therefore$ A's probability of winning $P(A) = \frac{1}{6} + \frac{5}{6} \cdot \frac{5}{6} \cdot \frac{1}{6} + \frac{5}{6} \cdot \frac{5}{6} \cdot \frac{5}{6} \cdot \frac{5}{6} \cdot \frac{1}{6} + \cdots$

This is a G.P. with $a = \frac{1}{6}$ and $r = \frac{5}{6} \cdot \frac{5}{6} = \frac{25}{36}$

$\therefore P(A) = \frac{a}{1-r} = \frac{\frac{1}{6}}{1-\frac{25}{36}} = \frac{6}{11}$, $P(B) = 1 - \frac{6}{11} = \frac{5}{11}$

## 2.3 Addition Law of Probability

**Statement**: If A and B be any two events, then

$P(A \cup B) = P(A) + P(B) - P(A \cap B)$

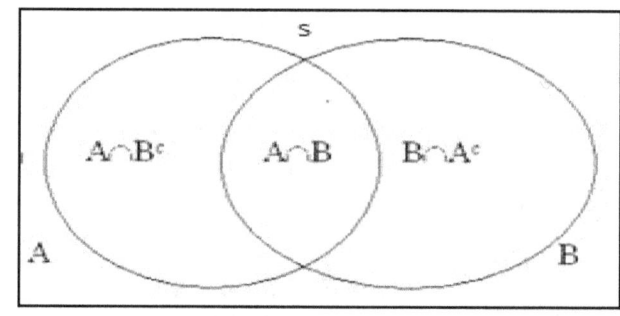

**Proof**: $A \cup B = A \cup (B \cap A^C)$

$\Rightarrow P(A \cup B) = P(A \cup (B \cap A^C))$

$\qquad = P(A) + P(B \cap A^C)$

($\because$ Both $A$ and $B \cap A^C$ are disjoint)

$\qquad = P(A) + [P(B \cap A^C) + P(A \cap B)] - P(A \cap B)$

$\qquad = P(A) + P[(B \cap A^C) \cup (A \cap B)] - P(A \cap B)$

$\qquad = P(A) + P(B) - P(A \cap B) \quad \because (B \cap A^C) \cup (A \cap B) = B$

Hence proved

➤ If A and B are mutually exclusive events, $A \cap B = \phi$

$\Rightarrow P(A \cup B) = P(A) + P(B)$

If $A, B, C$ are 3 events, then

$P(A \cup B \cup C) = P(A) + P(B) + P(C) - P(A \cap B) - P(B \cap C) - P(C \cap A) + P(A \cap B \cap C)$

**Example 13** Find the probability of getting a spade or an ace when a card is drawn from a well shuffled pack of 52 cards.

**Solution**: Let A: Getting a spade, $P(A) = \frac{13}{52}$

B: Getting an ace, $P(B) = \frac{4}{52}$

$A \cap B$: Getting an ace of spade, $P(A \cap B) = \frac{1}{52}$

Now $P(A \cup B) = P(A) + P(B) - P(A \cap B)$
$$= \frac{13}{52} + \frac{4}{52} - \frac{1}{52} = \frac{16}{52} = \frac{4}{13}$$

**Example 14** Find the probability of getting neither heart nor king when a card is drawn from a well shuffled pack of 52 cards.

**Solution**: Let $A$: Getting a card of heart, $P(A) = \frac{13}{52}$

$B$: Getting a card of king, $P(B) = \frac{4}{52}$

$A \cap B$: Getting a king of heart, $P(A \cap B) = \frac{1}{52}$

Now $P(A \cup B) = P(A) + P(B) - P(A \cap B)$
$$= \frac{13}{52} + \frac{4}{52} - \frac{1}{52} = \frac{16}{52} = \frac{4}{13}$$

Again Probability of neither heart nor king is given by $P(A^C \cap B^C)$

$P(A^C \cap B^C) = P(A \cup B)^C = 1 - P(A \cup B)$
$$= 1 - \frac{4}{13} = \frac{9}{13}$$

**Example 15** Three newspapers A, B, C are published in a city and a survey on readers reveals the following information: 25% read A, 30% read B, 20% read C, 10% read both A and B, 5% read both A and C, 8% read both B and C, 3% read all three newspapers. For a person chosen at random, find the probability that he reads none of the newspapers.

**Solution**: $P(A) = \frac{25}{100}$, $P(B) = \frac{30}{100}$, $P(C) = \frac{20}{100}$, $P(A \cap B) = \frac{10}{100}$,

$P(A \cap C) = \frac{5}{100}$, $P(B \cap C) = \frac{8}{100}$, $P(A \cap B \cap C) = \frac{3}{100}$

Now $P(A \cup B \cup C) = P(A) + P(B) + P(C) - P(A \cap B) - P(B \cap C) -$
$$P(C \cap A) + P(A \cap B \cap C)$$

$\Rightarrow P(A \cup B \cup C) = \frac{25}{100} + \frac{30}{100} + \frac{20}{100} - \frac{10}{100} - \frac{8}{100} - \frac{5}{100} + \frac{3}{100}$
$$= \frac{55}{100} = \frac{11}{20}$$

$\Rightarrow P(A \cup B \cup C)^C = 1 - P(A \cup B \cup C)$
$$= 1 - \frac{11}{20} = \frac{9}{20}$$

**Example 16** Discuss and comment on following:

$P(A) = \frac{1}{4}$, $P(B) = \frac{1}{3}$, $P(C) = \frac{2}{3}$ are probabilities of three mutually exclusive events $A, B, C$.

**Solution**: Since $A, B, C$ are mutually exclusive events,
$$P(A \cup B \cup C) = P(A) + P(B) + P(C)$$
$$\Rightarrow P(A \cup B \cup C) = \frac{1}{4} + \frac{1}{3} + \frac{2}{3} = \frac{5}{4} > 1$$

Which is not possible, hence it is a false statement.

## 2.4 Conditional Probability

The probability of occurrence of an event $A$, when event $B$ has already occurred is called conditional probability of $A$ and is denoted by $P(A|B)$.

If $P(A|B) = P(A)$, then event $A$ is said to be independent of event $B$.

### 2.4.1 Multiplicative Law of Probability

**Statement**: The probability of simultaneous occurrence of two events is equal to the probability of event multiplied by conditional probability of the other, i.e. for two events $A$ and $B$:
$$P(A \cap B) = P(A) \cdot P(B|A)$$

**Proof**: Suppose a trial results in $n$ exhaustive, mutually exclusive and equally likely outcomes, $m$ of them being favourable to the happening of event $A$,

then $P(A) = \dfrac{m}{n}$ \hspace{1cm} ...①

Out of $m$ outcomes, favorable to the happening of event $A$, let $m_1$ be favourable to happening of event $B$,

∴ Conditional probability of $B$ given that $A$ has already occurred is

$P(B|A) = \dfrac{m_1}{m}$ \hspace{1cm} ...②

Again out of $n$ exhaustive, mutually exclusive and equally likely outcomes, $m_1$ are favorable to happening of $A$ and $B$

∴ Probability of simultaneous occurrence of $A$ and $B$

$$= P(A \cap B) = \dfrac{m_1}{n}$$
$$= \dfrac{m_1}{n} \cdot \dfrac{m}{m} = \dfrac{m}{n} \cdot \dfrac{m_1}{m} = P(A) \cdot P(B|A) \quad \text{Using ① and ②}$$

∴ $P(A \cap B) = P(A) \cdot P\left(\dfrac{B}{A}\right)$

- $P(A \cap B)$ is also written as $P(AB)$
- Interchanging $A$ and $B$, $P(BA) = P(B) \cdot P(A|B)$
  
  or $P(AB) = P(B) \cdot P(A|B)$ \hspace{1cm} $\because A \cap B = B \cap A$

- If $A$ and $B$ are independent events, then $P(B|A) = P(B)$

  ∴ $P(AB) = P(A) \cdot P(B)$, if events $A$ and $B$ are independent

  Formulae for conditional probability:
  $$P(B|A) = \dfrac{P(A \cap B)}{P(A)}$$
  $$P(A|B) = \dfrac{P(A \cap B)}{P(B)}$$

**Example 17** An unbiased coin is tossed thrice. In which of the following cases events A and B are independent

(i)    $A$: The first throw results in a tail
       $B$: The last throw results in a head

(ii)    $A$: The number of tails is two
       $B$: The last throw results in a tail

**Solution**: $S = \{HHH, HHT, HTH, HTT, TTT, TTH, THT, THH\}$

(i) $A = \{TTT, TTH, THT, THH\}$, $P(A) = \frac{4}{8} = \frac{1}{2}$

$B = \{HHH, HTH, TTH, THH\}$, $P(B) = \frac{4}{8} = \frac{1}{2}$

$A \cap B = \{TTH, THH\}$, $P(A \cap B) = \frac{2}{8} = \frac{1}{4}$

Now $P(A \cap B) = \frac{1}{4}$, and $P(A).P(B) = \frac{1}{4}$

$\therefore P(A \cap B) = P(A).P(B)$

Hence events $A$ and $B$ are independent.

(ii) $A = \{HTT, THT, TTH\}$, $P(A) = \frac{3}{8}$

$B = \{HHT, HTT, TTT, THT\}$, $P(B) = \frac{4}{8} = \frac{1}{2}$

$A \cap B = \{HTT, THT\}$, $P(A \cap B) = \frac{2}{8} = \frac{1}{4}$

Now $P(A \cap B) = \frac{1}{4}$, and $P(A).P(B) = \frac{3}{16}$

$\therefore P(A \cap B) \neq P(A).P(B)$

Hence the events $A$ and $B$ are not independent.

**Example 18** If $P(A) = 0.3$, $P(B) = 0.5$ and $P(A|B) = 0.4$

Find (i) $P(A \cap B)$, (ii) $P(B|A)$, (iii) $P(A^C \cup B^C)$

**Solution**: (i) $P(A \cap B) = P(B).P\left(\frac{A}{B}\right) = 0.5 \times 0.4 = 0.2$

(ii) $P(B|A) = \frac{P(A \cap B)}{P(A)} = \frac{0.2}{0.3} = \frac{2}{3}$

(iii) $P(A^C \cup B^C) = P(A \cap B)^C$

$= 1 - P(A \cap B) = 1 - 0.2 = 0.8$

**Example 19** Cards are dealt one by one from a well shuffled pack of 52 cards until an ace appears. Show that the probability that exactly $n$ cards are dealt before the first ace appears is $\frac{4(51-n)(50-n)(49-n)}{52.51.50.49}$

**Solution**: Let $A$: Drawing $n$ non ace cards

$B$: Drawing an ace in $(n+1)^{th}$ draw

$P(A) = \frac{^{48}C_n}{^{52}C_n} = \frac{48!}{n!(48-n)!} \cdot \frac{n!(52-n)!}{52!}$

$= \frac{(52-n)(51-n)(50-n)(49-n)}{52.51.50.49}$

Also $P(B|A) = \frac{^4C_1}{(52-n)C_1} = \frac{4}{52-n}$

$$\therefore P(A \cap B) = P(A).P(B|A) = \frac{4(51-n)(50-n)(49-n)}{52.51.50.49}$$

**Example20** Two dice are thrown and sum of numbers appearing is observed to be 6. Find the conditional probability that number 2 has appeared at least once.

**Solution:** Let $A$: Number 2 has appeared at least once

$B$: Sum of numbers on two dice is 6

Now $P(A|B) = \frac{P(A \cap B)}{P(B)}$

$A \equiv \{(1,2),(2,2),(3,2),(4,2),(5,2),(6,2),(2,1),(2,3),(2,4),(2,5),(2,6)\}$

$B \equiv \{(1,5),(2,4),(3,3),(4,2),(5,1)\}$

$A \cap B \equiv \{(2,4),(4,2)\}$

Now $P(A \cap B) = \frac{2}{36}$ and $P(B) = \frac{5}{36}$ $\therefore P(A|B) = \frac{2}{5}$

**Example21** A bag contains 7 red 5 black balls; another bag contains 5 red and 8 black balls. A ball is drawn from the first bag and without noticing its colour is put in the second bag and then a ball is drawn from second bag. Find the probability that ball drawn is red in colour.

**Solution:** Case 1    $A$: Red ball is drawn from first bag.

$B$: Red ball is drawn from second bag.

$$P(A \cap B) = P(A).P\left(\frac{B}{A}\right)$$

$$= \frac{7}{12} \cdot \frac{6}{14} = \frac{42}{168}$$

Case 2    $C$: Black ball is drawn from first bag.

$B$: Red ball is drawn from second bag.

Then $P(C \cap B) = P(C).P\left(\frac{B}{C}\right)$

$$= \frac{5}{12} \cdot \frac{5}{14} = \frac{25}{168}$$

Required probability $= P(A \cap B) + P(C \cap B)$

$$= \frac{42}{168} + \frac{25}{168} = \frac{67}{168}$$

## 2.5 Bay's Theorem

If $E_1, E_2, \ldots, E_n$ are $n$ mutually exclusive and exhaustive events in a sample space such that $P(E_i) > 0$ for each $i, (i = 1,2,\ldots,n)$ and $A$ is an arbitrary event for which $P(A) \neq 0$, then the conditional probability of occurrence of $E_i$, given that $A$ has already occurred is given by

$$P(E_i/A) = \frac{P(E_i)P(A/E_i)}{\sum P(E_i)P(A/E_i)}, \quad i = 1,2,\ldots,n$$

**Proof:** As $E_1, E_2, \ldots, E_n$ are $n$ mutually exclusive and exhaustive events

$\therefore E_1 \cup E_2 \cup \ldots \cup E_n = S$

Now $A = A \cap S = A \cap (E_1 \cup E_2 \cup \ldots \cup E_n)$

$\Rightarrow A = (A \cap E_1) \cup (A \cap E_2) \cup (A \cap E_3) \cup \ldots \cup (A \cap E_n)$

By addition law of probability

$P(A) = P(A \cap E_1) + P(A \cap E_2) + P(A \cap E_3) + \cdots + P(A \cap E_n)$

$\because (A \cap E_1), (A \cap E_2), (A \cap E_3)$ are all mutually exclusive

$\Rightarrow P(A) = P(E_1)P(A/E_1) + P(E_2)P(A/E_2) + P(E_3)P(A/E_3) + \cdots + P(E_n)P(A/E_n)$

$= \sum P(E_i)P(A/E_i), \quad i = 1, 2, \ldots, n \qquad \ldots (1)$

$\therefore P(E_i/A) = \dfrac{P(E_i \cap A)}{P(A)} = \dfrac{P(E_i)P(A/E_i)}{\sum P(E_i)P(A/E_i)} \qquad \text{using } (1)$

**Example 22** Three urns I, II, III contain 6 red and 4 black balls, 2 red and 6 black balls, and 1 red and 8 black balls respectively. An urn is chosen randomly and a ball is drawn from the urn. If the ball drawn is red, find the probability that it was drawn from urn I.

**Solution:** Let $A$: A Red ball is drawn

$E_1$: Urn I is chosen, $P(E_1) = \dfrac{1}{3}$, $P(A/E_1) = \dfrac{6}{10} = \dfrac{3}{5}$

$E_2$: Urn II is chosen, $P(E_2) = \dfrac{1}{3}$, $P(A/E_2) = \dfrac{2}{8} = \dfrac{1}{4}$

$E_3$: Urn III is chosen, $P(E_3) = \dfrac{1}{3}$, $P(A/E_3) = \dfrac{1}{9}$

Required Probability $= P(E_1/A) = \dfrac{P(E_1)P(A/E_1)}{P(E_1)P(A/E_1)+P(E_2)P(A/E_2)+P(E_3)P(A/E_3)}$

$= \dfrac{\frac{1}{3} \cdot \frac{3}{5}}{\left(\frac{1}{3} \cdot \frac{3}{5}\right)+\left(\frac{1}{3} \cdot \frac{1}{4}\right)+\left(\frac{1}{3} \cdot \frac{1}{9}\right)} = \dfrac{108}{173}$

**Example 23** Two urns I and II contain 3 red and 4 black balls, 2 red and 5 black balls respectively. A ball is transferred from urn I to urn II and then a ball is drawn from urn II. If the ball drawn is found to be red, find the probability that the ball transferred from urn I is red.

**Solution:** Let $A$: A Red ball is drawn from urn II

$E_1$: Ball transferred from urn I is red, $P(E_1) = \dfrac{3}{7}$, $P(A/E_1) = \dfrac{3}{8}$

$E_2$: Ball transferred from urn I is black, $P(E_2) = \dfrac{4}{7}$, $P(A/E_2) = \dfrac{2}{8}$

Required Probability $= P(E_1/A) = \dfrac{P(E_1)P(A/E_1)}{P(E_1)P(A/E_1)+P(E_2)P(A/E_2)}$

$= \dfrac{\frac{3}{7} \cdot \frac{3}{8}}{\left(\frac{3}{7} \cdot \frac{3}{8}\right)+\left(\frac{4}{7} \cdot \frac{2}{8}\right)} = \dfrac{9}{9+8} = \dfrac{9}{17}$

**Example 24** An insurance company insured 2000 scooter drivers, 4000 car drivers and 6000 truck drivers. The probability of an accident involving a scooter driver is 0.01, a car driver is 0.03 and a truck driver is 0.15. If one of an insured person meets with an accident, what is the probability that he is a car driver?

**Solution**: Let $A$: An insured person meets with an accident

$E_1$: Person is an insured scooter driver

$$P(E_1) = \frac{2000}{2000+4000+6000} = \frac{2}{12}, \; P(A/E_1) = 0.01$$

$E_2$: Person is an insured car driver

$$P(E_2) = \frac{4000}{2000+4000+6000} = \frac{4}{12}, \; P(A/E_2) = 0.03$$

$E_3$: Person is an insured truck driver

$$P(E_3) = \frac{6000}{2000+4000+6000} = \frac{6}{12}, \; P(A/E_3) = 0.15$$

Required Probability $= P(E_2/A) = \dfrac{P(E_2)P(A/E_1)}{P(E_1)P(A/E_1)+P(E_2)P(A/E_2)+P(E_3)P(A/E_3)}$

$$= \frac{\frac{4}{12}(0.03)}{\frac{2}{12}(0.01)+\frac{4}{12}(0.03)+\frac{6}{12}(0.15)} = \frac{3}{26}$$

**Example 25** Ram speaks truth 2 out of 3 times and Shyam 4 out of 5 times; they agree in an assertion that from a bag containing 6 balls of different colour, a red ball has been drawn. Find the probability that the statement is true.

**Solution**: Let $A$: Ram and Shyam agree in the assertion that a red ball has been drawn

$E_1$: Red ball is drawn. $P(E_1) = \frac{1}{6}$, $P(A/E_1) = \frac{2}{3} \cdot \frac{4}{5}$

$E_2$: Non Red ball is drawn. $P(E_1) = \frac{5}{6}$, $P(A/E_2) = \frac{1}{3} \cdot \frac{1}{5}$

Required Probability $= P(E_1/A) = \dfrac{P(E_1)P(A/E_1)}{P(E_1)P(A/E_1)+P(E_2)P(A/E_2)}$

$$= \frac{\frac{1}{6} \cdot \frac{2}{3} \cdot \frac{4}{5}}{\left(\frac{1}{6} \cdot \frac{2}{3} \cdot \frac{4}{5}\right)+\left(\frac{5}{6} \cdot \frac{1}{3} \cdot \frac{1}{5}\right)} = \frac{8}{8+5} = \frac{8}{13}$$

**Example 26** A card from a deck of 52 cards is missing. 2 cards are drawn from the remaining deck of 51 cards and are found to be of spade. Find the probability that missing card is of spade.

**Solution**: Let $A$: 2 cards of spade are drawn from a deck of 51 cards.

$E_1$: Missing card is of spade, $\quad P(E_1) = \frac{13}{52}, \; P(A/E_1) = \frac{^{12}C_2}{^{51}C_2} = \frac{12 \times 11}{51 \times 50}$

$E_2$: Missing card is a non-spade, $P(E_2) = \frac{39}{52}, \; P(A/E_2) = \frac{^{13}C_2}{^{51}C_2} = \frac{13 \times 12}{51 \times 50}$

$$\therefore P(E_1/A) = \frac{P(E_1)P(A/E_1)}{P(E_1)P(A/E_1)+P(E_2)P(A/E_2)} = \frac{\frac{13 \times 12 \times 11}{52 \times 51 \times 50}}{\frac{13 \times 12 \times 11}{52 \times 51 \times 50}+\frac{39 \times 13 \times 12}{52 \times 51 \times 50}} = \frac{11}{11+39} = \frac{11}{50}$$

## 2.6 Random Variable and Probability Distributions

A random variable is often described as a variable whose values are determined by chance. The values taken by the variable may be countable or uncountable, based on which it is classified as discrete or continuous. Random variable is typically denoted using capital letters such as '$X$'.

### 2.6.1 Discrete Probability Distributions

A discrete probability distribution '$P(X)$' describes the probability of occurrence of each value of a discrete random variable '$X$'. A discrete random variable is a random variable that has countable values, such as a set of positive integers.

The discrete probability distribution '$P(X)$' of an experiment provides the corresponding probabilities to all possible values of the random variable '$X$' associated with it such that $\sum P(X) = 1$

**Example 27** Find the probability distribution of the number of aces when two cards are drawn at random with replacement from a well shuffled pack of 52 cards.

**Solution**: Let $X$ be a random variable showing number of aces. Clearly $X$ can take values 0, 1 or 2. If $S$ denotes success i.e. getting an ace and $F$ denotes failure i.e. getting a non-ace card, then $P(S) = \frac{4}{52} = \frac{1}{13}$ and $P(F) = \frac{12}{13}$

| X | Event | P(X) |
|---|-------|------|
| 0 | FF | $\frac{12}{13} \cdot \frac{12}{13} = \frac{144}{169}$ |
| 1 | SF or FS | $\left(\frac{1}{13} \cdot \frac{12}{13}\right) + \left(\frac{12}{13} \cdot \frac{1}{13}\right) = \frac{24}{169}$ |
| 2 | SS | $\frac{1}{13} \cdot \frac{1}{13} = \frac{1}{169}$ |

**Example 28** Find the probability distribution of the number of successes in two tosses of a die, when success is defined as a number greater than 4.

**Solution**: Let $X$ be a random variable showing number of successes in two tosses of a die. Clearly $X$ can take values 0, 1 or 2. If $S$ denotes success i.e. getting a number greater than 4 and $F$ denotes failure, then $P(S) = \frac{2}{6} = \frac{1}{3}$ and $P(F) = \frac{2}{3}$

| X | Event | P(X) |
|---|-------|------|
| 0 | FF | $\frac{2}{3} \cdot \frac{2}{3} = \frac{4}{9}$ |
| 1 | SF or FS | $\left(\frac{1}{3} \cdot \frac{2}{3}\right) + \left(\frac{2}{3} \cdot \frac{1}{3}\right) = \frac{4}{9}$ |
| 2 | SS | $\frac{1}{3} \cdot \frac{1}{3} = \frac{1}{9}$ |

**Example 29** 3 bad articles are mixed with 7 good ones. Find the probability distribution of number of bad articles if 3 are drawn at random without replacement from the lot.

**Solution**: Let $X$ be a random variable showing number of bad articles. Clearly $X$ can take values $0, 1, 2$ or $3$.

| X | Event | P(X) |
|---|---|---|
| 0 | 0 Bad 3 Good | $\dfrac{7_{C_3}}{10_{C_3}} = \dfrac{210}{720}$ |
| 1 | 1 Bad 2 Good | $\dfrac{3_{C_1} \cdot 7_{C_2}}{10_{C_3}} = \dfrac{378}{720}$ |
| 2 | 2 Bad 1 Good | $\dfrac{3_{C_2} \cdot 7_{C_1}}{10_{C_3}} = \dfrac{126}{720}$ |
| 3 | 3 Bad 0 Good | $\dfrac{3_{C_3}}{10_{C_3}} = \dfrac{6}{720}$ |

Note: Combination is being used as articles are drawn without replacement.

## 2.6.2 Mathematical Expectation

If $X$ be a random variable which can assume any one of the values $x_1, x_2, \cdots, x_n$ with respective probabilities $p_1, p_2, \cdots, p_n$; then the mathematical expectation of $X$ usually called as expected value of $X$, denoted by $E(X)$ is defined as:

$$E(X) = p_1 x_1 + p_2 x_2 + \cdots + p_n x_n = \sum p_i x_i \text{ ; where } \sum p_i = 1$$

**Physical interpretation of $E(X)$**

If $\bar{x}$ denotes mean of set of observations $x_1, x_2, \cdots, x_n$ with respective frequencies $f_1, f_2, \cdots, f_n$; then $\bar{x} = \dfrac{\sum f_i x_i}{N}, N = \sum f_i$

$$\Rightarrow \bar{x} = \dfrac{f_1}{N} x_1 + \dfrac{f_2}{N} x_2 + \cdots + \dfrac{f_n}{N} x_n \qquad \cdots \text{①}$$

Now out of total $N$ cases, $f_i$ are favorable to $x_i$

$$\therefore P(X = x_i) = \dfrac{f_i}{N} = p_i \text{ , } i = 1, 2, \cdots, n$$

$$\Rightarrow \dfrac{f_1}{N} = p_1, \dfrac{f_2}{N} = p_2, \cdots, \dfrac{f_n}{N} = p_n \qquad \cdots \text{②}$$

$$\therefore \bar{x} = p_1 x_1 + p_2 x_2 + \cdots + p_n x_n = \sum p_i x_i \qquad \text{using ② in ①}$$

Hence $\bar{x} = E(X)$

Hence mathematical expectation of a random variable is nothing but its arithmetic mean.

$\therefore$ We conclude Mean $(\bar{x}) = E(X) = \sum p_i x_i$

Similarly Variance $(\sigma^2) = \dfrac{\sum f_i x_i^2}{N} - \left(\dfrac{\sum f_i x_i}{N}\right)^2$

$$= \sum p_i x_i^2 - (\sum p_i x_i)^2 = E(X^2) - (E(X))^2$$

**Example 30** What is the expected number of heads appearing when a fair coin is tossed 3 times.

**Solution**: Let $X$ be a random variable showing number of heads. Clearly $X$ can take values $0, 1, 2, 3$.

| $x_i$ | Event | $p_i$ | $p_i x_i$ |
|---|---|---|---|
| 0 | TTT | $\frac{1}{8}$ | 0 |
| 1 | HTT, THT, TTH | $\frac{3}{8}$ | $\frac{3}{8}$ |
| 2 | HHT, HTH, THH | $\frac{3}{8}$ | $\frac{6}{8}$ |
| 3 | HHH | $\frac{1}{8}$ | $\frac{3}{8}$ |

$\therefore E(X) = \sum p_i x_i = \frac{12}{8} = 1.5$

**Example 31** A man draws 2 balls from a bag containing 3 white and 5 black balls. If he receives Rs 70 for every white ball he draws and Rs. 35 for every black ball, what is his expectation?

**Solution**: Following table shows the amount received by the man for each event:

| Event | Probability ($p_i$) | Amount ($x_i$) | $p_i x_i$ |
|---|---|---|---|
| 2 black balls | $\frac{5_{C_2}}{8_{C_2}} = \frac{10}{28}$ | $35+35 = 70$ | $\frac{10}{28} \times 70$ |
| 1 white 1 black ball | $\frac{3_{C_1} \times 5_{C_1}}{8_{C_2}} = \frac{15}{28}$ | $70+35 = 105$ | $\frac{15}{28} \times 105$ |
| 2 white balls | $\frac{3_{C_2}}{8_{C_2}} = \frac{3}{28}$ | $70+70 = 140$ | $\frac{3}{28} \times 140$ |

$\therefore E(X) = \sum p_i x_i = \frac{700}{28} + \frac{1575}{28} + \frac{420}{28} = \frac{385}{4} = 96.25$

**Example 32** For a random variable $X$, the probability mass function is

$f(x) = kx$, for $x = 1, 2, \cdots, n$

$\quad\quad\quad = 0$, otherwise

Find expectation of $X$.

**Solution**: Here $f(x)$ denotes probability mass function

$\therefore \sum_{x=1}^{n} f(x) = \sum_{x=1}^{n} kx = k \sum_{x=1}^{n} x = 1$

$\Rightarrow k \frac{n(n+1)}{2} = 1$

$\Rightarrow k = \frac{2}{n(n+1)}$

$$\therefore E(X) = \sum_{x=1}^{n} xf(x) = \sum_{x=1}^{n} xkx = \sum_{x=1}^{n} x^2 \frac{2}{n(n+1)} = \frac{2}{n(n+1)} \sum_{x=1}^{n} x^2$$

$$= \frac{2}{n(n+1)}(1^2 + 2^2 + \cdots + n^2) = \frac{2}{n(n+1)} \frac{n(n+1)(2n+1)}{6} = \frac{(2n+1)}{3}$$

**Example 33** A random variable $X$ has the following probability function:

| X | -2 | -1 | 0 | 1 | 2 | 3 |
|---|----|----|---|---|---|---|
| P(X) | k | 0.1 | 0.3 | 2k | 0.2 | K |

Calculate mean and variance.

**Solution**: For any probability function $\sum p_i = 1$

$\Rightarrow k + 0.1 + 0.3 + 2k + 0.2 + k = 1$

$\Rightarrow k = 0.1$

| $x_i$ | $p_i$ | $p_i x_i$ | $p_i x_i^2$ |
|---|---|---|---|
| -2 | 0.1 | -0.2 | 0.4 |
| -1 | 0.1 | -0.1 | 0.1 |
| 0 | 0.3 | 0 | 0 |
| 1 | 0.2 | 0.2 | 0.2 |
| 2 | 0.2 | 0.4 | 0.8 |
| 3 | 0.1 | 0.3 | 0.9 |
|  |  | $\sum p_i x_i = 0.6$ | $\sum p_i x_i^2 = 2.4$ |

$\therefore$ Mean $= \sum p_i x_i = 0.6$

Variance $(\sigma^2) = \sum p_i x_i^2 - (\sum p_i x_i)^2$

$= 2.4 - 0.36 = 2.04$

## 2.6.2 Continuous Probability Distributions

The probability distribution $P(X)$ associated with a continuous random variable $X$ is called a continuous distribution. A continuous random variable is having a set of infinite and uncountable values, for example set of real numbers in the interval $(0,1)$ is uncountable.

If $X$ be a continuous random variable taking values in the interval $[a,b]$, the function $f(x)$ is said to be the Probability Density Function (PDF) of $X$, if it satisfies the following properties:

i.  $f(x) \geq 0 \quad \forall x \in X$ in $[a,b]$.
ii. Total area under the probability curve is one, i.e. $P(a \leq x \leq b) = 1$.

For two distinct points $c$ and $d$ in the interval $[a,b]$; $P(c \leq x \leq d)$ is given by area under the probability curve between the ordinates $x = c$ and $x = d$, i.e. $P(c \leq x \leq d) = \int_c^d f(x)dx$.

Also $F(x) = P(X \leq x) = \int_{-\infty}^x f(x)dx$ is called cumulative distribution function or simply distribution function.

**Example 34** Find whether the following is a probability density function:
$$f(x) = \begin{cases} x, & 0 \leq x \leq 1 \\ 2x, & 1 < x \leq 2 \end{cases}$$

**Solution:** For $f(x)$ to be a probability density function, $\int_{-\infty}^{\infty} f(x)\, dx = 1$

$$\int_{-\infty}^{\infty} f(x)\, dx = \int_{-\infty}^{0} f(x)\, dx + \int_{0}^{1} f(x)\, dx + \int_{1}^{2} f(x)\, dx + \int_{2}^{\infty} f(x)\, dx$$

$$= \int_{-\infty}^{0} 0\, dx + \int_{0}^{1} x\, dx + \int_{1}^{2} 2x\, dx + \int_{2}^{\infty} 0\, dx$$

$$= 0 + \frac{1}{2}[x^2]_0^1 + [x^2]_1^2$$

$$= \frac{1}{2}[1-0] + [4-1] = \frac{7}{2} \neq 1$$

Hence $f(x)$ is not a probability density function.

**Example 35** Let $X$ be a random variable with PDF given by
$$f(x) = kx^4, \ |x| \leq 1$$
$$= 0, \text{ otherwise}$$

i. Find the value of the constant $k$
ii. Find $E(X)$ and $Var(X)$
iii. Find $P(X) \geq \frac{1}{2}$

**Solution:** i. For $f(x)$ to be PDF, $\int_{-\infty}^{\infty} f(x)\, dx = 1$

$$\Rightarrow \int_{-1}^{1} kx^4\, dx = 1$$

$$\Rightarrow \frac{k}{5}[x^5]_{-1}^{1} = 1 \quad \Rightarrow k = \frac{5}{2}$$

ii. $E(X) = \int_{-\infty}^{\infty} x f(x)\, dx$

$$= \int_{-1}^{1} x k x^4\, dx$$

$$= k \int_{-1}^{1} x^5\, dx = 0 \quad (\because x^5 \text{ is an odd function})$$

Also $Var(X) = E(X^2) - (E(X))^2$

$$= \int_{-\infty}^{\infty} x^2 f(x)\, dx - 0$$

$$= k \int_{-1}^{1} x^6\, dx = \frac{5}{7}$$

iii. $P(X) \geq \frac{1}{2} = \int_{\frac{1}{2}}^{1} kx^4\, dx = \frac{k}{5}[x^5]_{\frac{1}{2}}^{1} = \frac{31}{64}$

**Example 36** Show that the function $f(x)$ defined as
$$f(x) = \begin{cases} e^{-x}, & x \geq 0 \\ 0, & x < 0 \end{cases}$$, is a probability density function and find the probability that the variate $X$ having $f(x)$ as density function will lie in the interval (1,2). Also find the probability distribution function $F(2)$.

**Solution**: We have $\int_{-\infty}^{\infty} f(x)dx = \int_{-\infty}^{0} 0\,dx + \int_{0}^{\infty} e^{-x}\,dx = -[e^{-x}]_0^{\infty} = 1$

$\therefore f(x)$ is a probability density function.

Also $P(1 \leq x \leq 2) = \int_{1}^{2} e^{-x}\,dx = -[e^{-x}]_1^2 = 2.33$

Again $F(2) = \int_{-\infty}^{2} f(x)dx = \int_{-\infty}^{0} 0\,dx + \int_{0}^{2} e^{-x}\,dx = -[e^{-x}]_0^2 = 1 - e^{-2}$

$= 1 - 0.135 = 0.865$

## 2.7 Moments

The expected values $E(x-a), E((x-a)^2), E((x-a)^3),\ldots, E((x-a)^r)$ are called moments about any point $a$.

Thus $r^{th}$ moment about any point '$a$' of any distribution is denoted by $\mu_r'$ and is given by: $\mu_r' = \sum p_i (x_i - a)^r = \frac{1}{N}\sum f_i (x_i - a)^r$, where $N = \sum f_i$

In particular $r^{th}$ moment about mean $\bar{x}$ is given by $\mu_r = \frac{1}{N}\sum f_i (x_i - \bar{x})^r$

**Some important results:**

- $\mu_0 = \mu_0' = 1$
- $\mu_1 = \frac{1}{N}\sum f_i (x_i - \bar{x}) = \frac{1}{N}\sum f_i x_i - \frac{\bar{x}}{N}\sum f_i = \bar{x} - \bar{x} = 0$
- $\mu_2 = \frac{1}{N}\sum f_i (x_i - \bar{x})^2 = \frac{1}{N}\sum f_i x_i^2 + \frac{\bar{x}^2}{N}\sum f_i - 2\frac{\bar{x}}{N}\sum f_i x_i$

$= \frac{1}{N}\sum f_i x_i^2 + \bar{x}^2 - 2\bar{x}^2 = \frac{1}{N}\sum f_i x_i^2 - \bar{x}^2 = \sigma^2$

- $\mu_1' = \frac{1}{N}\sum f_i (x_i - a) = \frac{1}{N}\sum f_i x_i - \frac{a}{N}\sum f_i = \bar{x} - a$

**Relation between $\mu_r'$ and $\mu_r$**

$\mu_r = \frac{1}{N}\sum f_i (x_i - \bar{x})^r$

$= \frac{1}{N}\sum f_i ((x_i - a) - (\bar{x} - a))^r$

$= \frac{1}{N}\sum f_i (d - \mu_1')^r$

by putting $x_i - a = d$

$\Rightarrow \mu_r = \frac{1}{N}\sum f_i \left(d^r - r_{C_1} d^{r-1}\mu_1' + r_{C_2} d^{r-2}\mu_1'^2 + \cdots + (-1)^r \mu_1'^r\right)$

$= \frac{1}{N}\sum f_i d^r - r_{C_1}\mu_1'\frac{1}{N}\sum f_i d^{r-1} + r_{C_2}\mu_1'^2\frac{1}{N}\sum f_i d^{r-2} + \cdots + (-1)^r \mu_1'^r \frac{1}{N}\sum f_i$

$\Rightarrow \mu_r = \mu_r' - r_{C_1}\mu_{r-1}'\mu_1' + r_{C_2}\mu_{r-2}'\mu_1'^2 + \cdots + (-1)^r \mu_1'^r$

In particular

$\mu_2 = \mu_2' - 2\mu_1'^2 + \mu_0'\mu_1'^2 = \mu_2' - \mu_1'^2$ as $\mu_0' = 1$

$\mu_3 = \mu_3' - 3\mu_2'\mu_1' + 3\mu_1'^3 - \mu_0'\mu_1'^3 = \mu_3' - 3\mu_2'\mu_1' + 2\mu_1'^3$

Similarly $\mu_4 = \mu_4' - 4\mu_3'\mu_1' + 6\mu_2'\mu_1'^2 - 3\mu_1'^4$

## 2.8 Moment Generating Function (MGF)

The moment generating function of the variable $x$ about a point $x = a$ is defined as the expected value of $e^{t(x-a)}$ and is denoted by $M_a(t)$.

$\therefore M_a(t) = E[e^{t(x-a)}]$.

$\Rightarrow M_a(t) = \sum p_i \, e^{t(x_i - a)}$ ...①

$= \sum p_i + t \sum p_i (x_i - a) + \dfrac{t^2}{2!} \sum p_i (x_i - a)^2 + \ldots + \dfrac{t^r}{r!} \sum p_i (x_i - a)^r + \cdots$

$(\because e^x = 1 + x + \dfrac{x^2}{2!} + \cdots + \dfrac{x^r}{r!} + \cdots)$

$\Rightarrow M_a(t) = 1 + t\mu_1' + \dfrac{t^2}{2!}\mu_2' + \cdots + \dfrac{t^r}{r!}\mu_r' + \cdots$ ...②

$\because p_i = \dfrac{f_i}{N}$

where $\mu_r'$ is the moment of order $r$ about a point $a$

Hence $\mu_r'$ = coefficient of $\dfrac{t^r}{r!}$

Thus $M_a(t)$ generates moments and therefore it is called moment generating function.

Again rewriting ① as $M_a(t) = e^{-at} \sum p_i \, e^{tx_i}$

$\Rightarrow M_a(t) = e^{-at} M_0(t)$

Thus (MGF about the point $a$) $= e^{-at}$ (MGF about origin)

Again if $f(y)$ be density function of a continuous variate $Y$, then the moment generating function of the continuous probability distribution about $y = a$ is given by:

$$M_a(t) = \int_{-\infty}^{\infty} e^{t(y-a)} f(y) \, dy$$

We can also generate moments by differentiating $r$ times w.r.t. $t$ and then putting $t = 0$

i.e. $\mu_r' = \left[\dfrac{d^r}{dt^r} M_a(t)\right]_{t=0}$ ...③

Thus the moments about any point $x = a$ can be found using ② or more conveniently using ③

If a moment-generating function exists for a random variable $X$, then:

- $M(0) = 1$
- $M'(0) = E(X)$, $M''(0) = E(X^2)$, $M'''(0) = E(X^3), \cdots$
- The mean and variance of $X$ can be found by evaluating the first and second derivatives of the moment-generating function at $t = 0$.
  i.e. $\bar{x} = E(X) = M'(0)$, $\sigma^2 = E(X^2) - (E(X))^2 = M''(0) - [M'(0)]^2$

**Example 37** Find the moment generating function for the probability distribution given by number of heads appearing when a fair coin is tossed 3 times and hence find mean and variance.

**Solution**: Let $X$ be a random variable showing number heads. Clearly $X$ can take values $0, 1, 2, 3$. Probability distribution is given by:

| $x_i$ | 0 | 1 | 2 | 3 |
|---|---|---|---|---|
| $p_i$ | $\frac{1}{8}$ | $\frac{3}{8}$ | $\frac{3}{8}$ | $\frac{1}{8}$ |

$M(t) = E(e^{tX}) = \sum p_i \, e^{tx_i}$

$\Rightarrow M(t) = \frac{1}{8} + \frac{3}{8}e^t + \frac{3}{8}e^{2t} + \frac{1}{8}e^{3t}$

$M'(t) = \frac{3}{8}e^t + \frac{6}{8}e^{2t} + \frac{3}{8}e^{3t}$

$\therefore$ Mean $= M'(0) = \frac{3}{8} + \frac{6}{8} + \frac{3}{8} = \frac{12}{8} = 1.5$

Also $M''(t) = \frac{3}{8}e^t + \frac{12}{8}e^{2t} + \frac{9}{8}e^{3t}$

$\Rightarrow M''(0) = \frac{3}{8} + \frac{12}{8} + \frac{9}{8} = \frac{24}{8} = 3$

$\therefore$ Variance $= M''(0) - [M'(0)]^2 = 3 - (1.5)^2 = 0.75$

**Example 38** Let $X$ be a random variable with PDF given by

$$f(x) = \begin{cases} ke^{-kx}, & x \in (0, \infty) \\ 0, & \text{otherwise} \end{cases}$$

Find moment generating function of $f(x)$, hence find mean and variance.

**Solution:** $M(t) = E(e^{tX}) = \int_{-\infty}^{\infty} e^{tx} f(x) \, dx$

$= \int_{-0}^{\infty} e^{tx} k e^{-kx} \, dx = k \int_{-0}^{\infty} e^{-(k-t)x} \, dx = k \left[ \frac{e^{-(k-t)}}{t-k} \right]_0^{\infty} = \frac{k}{k-t}$

Also $M'(t) = \frac{k}{(k-t)^2}$

$\therefore$ Mean $= M'(0) = \frac{1}{k}$

Also $M''(t) = \frac{2k}{(k-t)^3} \Rightarrow M''(0) = \frac{2}{k^2}$

$\therefore$ Variance $= M''(0) - [M'(0)]^2 = \frac{2}{k^2} - \frac{1}{k^2} = \frac{1}{k^2}$

## 2.9 Skewness

Skewness is a measure of the asymmetry of the probability distribution of a random variable about its mean. In a symmetrical distribution, mean, mode and median coincide.
The skewness value can be positive or negative.

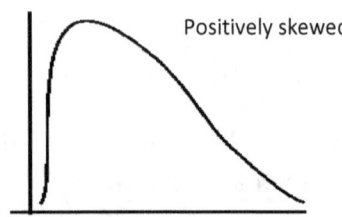

Positively skewed

Mass of distribution is concentrated to left

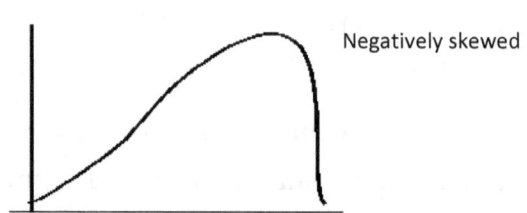

Negatively skewed

Mass of distribution is concentrated to the right

Coefficient of skewness $= \dfrac{\text{Mean} - \text{Mode}}{\text{Standard Deviation}}$

∴ $S_k = \frac{M-M_o}{\sigma}$, $S_k$ is called Karl Pearson's coefficient of skewness and lies between $-3$ and $+3$.

If mode is ill defined, then using $M_o = 3M_d - 2M$

$S_k = \frac{3(M-M_d)}{\sigma}$

Karl Pearson defined the following four coefficients based upon the first four moments about the mean:

$$\beta_1 = \frac{\mu_3^2}{\mu_2^3}, \quad \gamma_1 = \sqrt{\beta_1}$$

$$\beta_2 = \frac{\mu_4}{\mu_2^2}, \quad \gamma_2 = \beta_2 - 3$$

$\beta_1$ gives a measure of departure from symmetry and $\beta_2$ is associated with skewness.

## 2.10 Kurtosis

Kurtosis measures the degree of peakness of a distribution and is given by $\beta_2$

If $\beta_2 > 3$, the curve is peaked or leptokurtic

If $\beta_2 = 3$, the curve is normal or mesokurtic

If $\beta_2 < 3$, the curve is flat topped or platykurtic

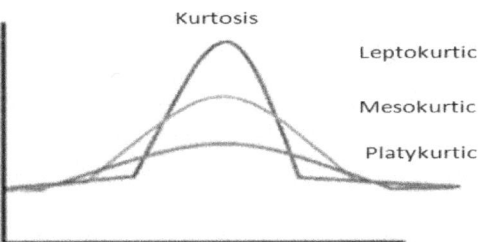

**Example39** Calculate the first four moments of the following distribution about the mean and hence find $\beta_1$ and $\beta_2$.

| x | 0 | 1 | 2 | 3 | 4 | 5 | 6 | 7 | 8 |
|---|---|---|---|---|---|---|---|---|---|
| f | 1 | 8 | 28 | 56 | 70 | 56 | 28 | 8 | 1 |

**Solution**: Taking 4 as assumed mean, let us first calculate moments about $x = 4$

$\mu_r' = \frac{1}{N}\Sigma f(x-a)^r = \frac{1}{N}\Sigma f d^r$, where $d = x - 4$

| x | f | $d = x-4$ | $fd$ | $fd^2$ | $fd^3$ | $fd^4$ |
|---|---|---|---|---|---|---|
| 0 | 1 | −4 | −4 | 16 | −64 | 256 |
| 1 | 8 | −3 | −24 | 72 | −216 | 648 |
| 2 | 28 | −2 | −56 | 112 | −224 | 448 |
| 3 | 56 | −1 | −56 | 56 | −56 | 56 |
| 4 | 70 | 0 | 0 | 0 | 0 | 0 |
| 5 | 56 | 1 | 56 | 56 | 56 | 56 |
| 6 | 28 | 2 | 56 | 112 | 224 | 448 |
| 7 | 8 | 3 | 24 | 72 | 216 | 648 |
| 8 | 1 | 4 | 4 | 16 | 64 | 256 |
| | N= 256 | | 0 | 512 | 0 | 2816 |

$\mu_1' = \frac{1}{N}\Sigma fd = 0$, $\quad \mu_2' = \frac{1}{N}\Sigma fd^2 = \frac{512}{256} = 2$,

$\mu_3' = \frac{1}{N}\Sigma fd^3 = 0$, $\quad \mu_4' = \frac{1}{N}\Sigma fd^4 = \frac{2816}{256} = 11$

Moments about mean:

$\mu_1 = 0$

$\mu_2 = \mu_2' - \mu_1'^2 = 2$

$\mu_3 = \mu_3' - 3\mu_2'\mu_1' + 2\mu_1'^3 = 0$

$\mu_4 = \mu_4' - 4\mu_3'\mu_1' + 6\mu_2'\mu_1'^2 - 3\mu_1'^4 = 11$

$\beta_1 = \dfrac{\mu_3^2}{\mu_2^3} = 0$, $\beta_2 = \dfrac{\mu_4}{\mu_2^2} = \dfrac{11}{4} = 2.75$

## Exercise 2

1. In a single throw of two dice, what is the probability of getting six as a product of numbers on two dice?
2. A bag contains 4 black, 2 red and 3 blue pens. If 2 pens are drawn at random from the pack and then another pen is drawn without replacement, what is the probability of drawing 2 black and 1 blue pens?
3. A purse contains 2 silver and 4 copper coins. A second purse contains 4 silver and 3 copper coins. If a coin is pulled out at random from one of the two purses, what is the probability that coin is a silver coin?
4. A problem in mathematics is given to three students A, B and C whose chances of solving are $\dfrac{1}{2}, \dfrac{1}{3}$ and $\dfrac{1}{4}$ respectively. What is the probability that the problem will not be solved?
5. If four whole numbers taken at random are multiplied together, find the probability that the last digit in the product is 1,3,7 or 9.
6. A speaks truth in 60% and B in 75% cases of the cases. In what percentage of cases they are likely to contradict each other in stating the same fact.
7. A and B take turns in throwing two dice and the first to throw 10 wins the game. If A has the first throw, find B's chances of winning.
8. A bag contains 10 white and 3 black balls. Another bag contains 3 white and 5 black balls. Two balls are transferred from the first bag and placed in the second and then one ball is drawn from the second bag. What is the probability that it is a white ball?
9. In each of a set of games it is 2 to 1 in favour the winner of the previous game. What is the probability that who wins the first game shall win at least 3 games out of the next four games?
10. A and B take turns in throwing two dice with A having the first trial. The first to throw 10 being awarded a prize. Find their expectations if the prize money is Rs 460.
11. In a lottery $m$ tickets are drawn out of $n$ tickets numbered from 1 to $n$. Show that the expectation of sum of numbers drawn is $m\left(\dfrac{n+1}{2}\right)$.
12. Find the probability distribution of number of kings drawn when two cards are drawn one by one, without replacement, from a deck of 52 cards.

13. A random variable X has the following probability distribution:

| X | 0 | 1 | 2 | 3 | 4 | 5 |
|---|---|---|---|---|---|---|
| P(X) | 0.1 | K | 0.2 | 2K | 0.3 | K |

    Find  i. K    ii. $P(X \leq 1)$    iii. $P(X > 3)$

14. Let X be a random variable with PDF given by

$$f(x) = \begin{cases} \frac{1}{4}, & |x| \leq 2 \\ 0, & \text{otherwise} \end{cases}$$

    Obtain the values of  i. $P(X < 1)$  ii. $P(|X| > 1)$  iii. $P(2X + 3) > 5)$

15. For the function $f(x)$ defined by $f(x) = ce^{-x}, 0 \leq x \leq \infty$, find the value of $c$ which changes $f(x)$ to a probability density function.

16. If X be a random variable with PDF given by

$$f(x) = \begin{cases} e^{-x}, & x \in (0, \infty) \\ 0, & \text{otherwise} \end{cases}, \text{ find moment generating function of } f(x).$$

17. Find $E(X^3)$ using the moment generating function $(1 - 2t)^{-10}$

## Answers

1. $\frac{1}{9}$
2. $\frac{1}{14}$
3. $\frac{19}{42}$
4. $\frac{1}{4}$
5. $\frac{16}{625}$
6. 45%
7. $\frac{11}{23}$
8. $\frac{59}{130}$
9. $\frac{4}{9}$
10. Rs.240, Rs. 220
12. $X = 0$; $P(X) = \frac{188}{221}$, $X = 1$; $P(X) = \frac{32}{221}$, $X = 2$; $P(X) = \frac{1}{221}$
13. i. 0.1    ii. 0.2    iii. 0.4
14. i. $\frac{3}{4}$  ii. $\frac{1}{2}$  iii. $\frac{1}{4}$
15. 1
16. $\frac{1}{1-t}$
17. 10,560.

*Chapter 3: Theoretical Probability Distributions*

# Theoretical Probability Distributions

## 3.1 Introduction

Probability distributions are either discrete or continuous, depending on whether they define probabilities for discrete or continuous variables. Here we shall confine our studies to Binomial, Poisson and Normal distributions of which Binomial and Poisson distributions are discrete distributions whereas Normal distribution is a continuous probability distribution.

## 3.2 Binomial Distribution

A series of independent trials which result in one of the two mutually exclusive outcomes 'success' or 'failure' such that the probability of the success (or failure) in each trials is constant, then such repeated independent trials are called as 'Bernoulli trials'. A discrete random variable which results in only one of the two possible outcomes (success or failure) is called Binomial variable.

Let there be $n$ independent finite trials in an experiment such that

i. Each trial has only two possible outcomes success and failure
ii. Probability of success $(p)$ and probability of failure $(q)$ are constant for all the trials and $p + q = 1$.

Then if a random variable $X$ denotes the number of successes in $n$ trials, then

$P(X = r) = {}^nC_r\, p^r q^{n-r}$ or $P(r) = {}^nC_r\, q^{n-r} p^r$

$\therefore$ Binomial distribution may be given as $(q + p)^r$

### 3.2.1 Mean of Binomial Distribution

Mean $= \sum_{r=0}^{n} r P(r)$

$= \sum_{r=0}^{n} r\, {}^nC_r\, q^{n-r} p^r$

$= {}^nC_1\, q^{n-1} p^1 + 2\,{}^nC_2\, q^{n-2} p^2 + 3\,{}^nC_3\, q^{n-3} p^3 + \cdots + n\,{}^nC_n\, q^0 p^n$

$= n\, q^{n-1} p + \frac{2n(n-1)}{2!}\, q^{n-2} p^2 + \frac{3n(n-1)(n-2)}{3!}\, q^{n-3} p^3 + \cdots + n p^n$

$= np\left[ q^{n-1} + (n-1)\, q^{n-2} p + \frac{(n-1)(n-2)}{2!}\, q^{n-3} p^2 + \cdots + p^{n-1} \right]$

$= np\left[ {}^{n-1}C_0\, q^{n-1} + {}^{n-1}C_1\, q^{n-2} p + {}^{n-1}C_2\, q^{n-3} p^2 + \cdots + {}^{n-1}C_{n-1}\, p^{n-1} \right]$

$= np\, (q + p)^{n-1}$

$= np \qquad \because q + p = 1$

### 3.2.2 Variance of Binomial Distribution

Variance $= \sum_{r=0}^{n} r^2 P(r) - (\text{mean})^2$

Now $\sum_{r=0}^{n} r^2 P(r) = \sum_{r=0}^{n} r^2\, {}^nC_r\, q^{n-r} p^r$

$$= {}^nC_1 q^{n-1}p + 2^2 \, {}^nC_2 q^{n-2}p^2 + 3^2 \, {}^nC_3 q^{n-3}p^3 + \cdots + n^2 \, {}^nC_n p^n$$

$$= n q^{n-1}p + \frac{4n(n-1)}{2!} q^{n-2}p^2 + \frac{9n(n-1)(n-2)}{3!} q^{n-3}p^3 + \cdots + n^2 p^n$$

$$= np \left[ q^{n-1} + 2(n-1) q^{n-2}p + \frac{3(n-1)(n-2)}{2!} q^{n-3}p^2 + \cdots + np^{n-1} \right]$$

$$= np \left[ q^{n-1} + (n-1)q^{n-2}p + \frac{(n-1)(n-2)}{2!} q^{n-3}p^2 + \cdots + p^{n-1} \right]$$

$$+ np \left[ (n-1)q^{n-2}p + (n-1)(n-2) q^{n-3}p^2 + \cdots + (n-1)p^{n-1} \right]$$

$$= np \left[ {}^{n-1}C_0 q^{n-1} + {}^{n-1}C_1 q^{n-2}p + {}^{n-1}C_2 q^{n-3}p^2 + \cdots + {}^{n-1}C_{n-1} p^{n-1} \right]$$

$$+ np(n-1)p \left[ {}^{n-2}C_0 q^{n-2} + {}^{n-2}C_1 q^{n-3}p + \cdots + {}^{n-2}C_{n-2} p^{n-2} \right]$$

$$= np[(q+p)^{n-1} + (n-1)p(q+p)^{n-2}]$$

$$= np[1 + (n-1)p] \qquad \because q + p = 1$$

$\therefore$ Variance $= np[1 + np - p] - n^2p^2 = np[q + np] - n^2p^2 \quad \because 1 - p = q$

$$= npq$$

$\therefore$ For Binomial distribution

Mean $= np$

$\mu_1 = 0$

$\mu_2 = \sigma^2 = npq \qquad \because \mu_2 = variance \, (\sigma^2)$

Similarly $\mu_3 = npq(q-p)$

$\mu_4 = npq[1 + 3pq(n-2)]$

$\beta_1 = \frac{\mu_3^2}{\mu_2^3} = \frac{[npq(q-p)]^2}{(npq)^3} = \frac{(q-p)^2}{npq}$, $\quad \gamma_1 = \frac{(q-p)}{\sqrt{npq}}$

$\beta_2 = \frac{\mu_4}{\mu_2^2} = \frac{1 + 3pq(n-2)}{npq} = 3 + \frac{1-6pq}{npq}$, $\quad \gamma_2 = \beta_2 - 3 = \frac{1-6pq}{npq}$

**Remarks**:

- Binomial Distribution is symmetrical if $\beta_1 = 0$ i.e. if $\frac{(q-p)^2}{npq} = 0$ or $p = q = \frac{1}{2}$
- Binomial Distribution is positively skewed if $\gamma_1 > 0$ i.e. if $q - p > 0$ or $1 - 2p > 0$ or $p < \frac{1}{2}$
- Binomial Distribution is negatively skewed if $p > \frac{1}{2}$
- Since $0 < q < 1$, $\therefore$ for binomial distribution $npq < np$ i.e. vaiance $<$ mean.

### 3.2.3 Moment Generating Function of Binomial Distribution

Moment Generating Function (MGF) about origin is expected value of $e^{tr}$

$M_0(t) = E(e^{tr})$

$= \sum p(r)e^{tr}$

$= \sum {}^nC_r p^r q^{n-r} e^{tr}$

$$= \sum {}^nC_r (pe^t)^r q^{n-r}$$
$$= (q + pe^t)^n$$

Now $\left[\frac{d}{dt} M_0(t)\right]_{t=0} = [n(q + pe^t)^{n-1} pe^t]_{t=0} = n(q + p)^{n-1} p = np$

$\Rightarrow \mu'_1 = np$

Again $M_a(t) = e^{-at} M_0(t)$

$\therefore M_m(t) = e^{-npt}(q + pe^t)^n$ where $m = np$ denotes mean of the distribution

$$= (qe^{-pt} + pe^{qt})^n$$

$$= \left[(q+p) + (-qpt + pqt) + \left(\frac{qp^2t^2}{2!} + \frac{pq^2t^2}{2!}\right) + \left(\frac{-qp^3t^3}{3!} + \frac{pq^3t^3}{3!}\right) + \left(\frac{qp^4t^4}{4!} + \frac{pq^4t^4}{4!}\right) + \cdots\right]^n$$

$$= \left[1 + \left(pq\frac{t^2}{2!} + pq(q^2 - p^2)\frac{t^3}{3!} + pq(q^3 + p^3)\frac{t^4}{4!} + \cdots\right)\right]^n$$

$$= 1 + n\left(pq\frac{t^2}{2!} + pq(q^2 - p^2)\frac{t^3}{3!} + pq(q^3 + p^3)\frac{t^4}{4!} + \cdots\right)$$

$$+ \frac{n(n-1)}{2!}\left(pq\frac{t^2}{2!} + pq(q^2 - p^2)\frac{t^3}{3!} + pq(q^3 + p^3)\frac{t^4}{4!} + \cdots\right)^2 + \cdots$$

$\therefore M_m(t) = 1 + npq\frac{t^2}{2!} + npq(q - p)\frac{t^3}{3!} +$

$\quad [npq(q^2 + p^2 - pq) + 3n(n-1)p^2q^2]\frac{t^4}{4!} + \cdots \quad \because (q+p) = 1$

Now $M_m(t) = 1 + \mu_1 t + \mu_2 \frac{t^2}{2!} + \mu_3 \frac{t^3}{3!} + \mu_4 \frac{t^4}{4!}$

$\therefore 1 + \mu_1 t + \mu_2 \frac{t^2}{2!} + \mu_3 \frac{t^3}{3!} + \mu_4 \frac{t^4}{4!} = 1 + npq\frac{t^2}{2!} + npq(q - p)\frac{t^3}{3!} +$

$\quad [npq(q^2 + p^2 - pq) + 3n(n-1)p^2q^2]\frac{t^4}{4!} + \cdots$

Comparing coefficients of different powers of $t$ on both sides, we get

$\mu_1 = 0$

$\mu_2 = npq$

$\mu_3 = npq(q - p)$

$\mu_4 = npq(q^2 + p^2 - pq) + 3n(n-1)p^2q^2$

$\quad = npq(1 - 3pq) + 3n(n-1)p^2q^2 \quad \because q^2 + p^2 = (q+p)^2 - 2pq$

$\quad = npq(1 - 3pq + 3npq - 3pq)$

$\quad = npq[1 + 3pq(n - 2)]$

**Example1** Show that Variance of a binomial distribution is less than or equal to $\frac{n}{4}$.

**Solution**: Variance ($\sigma^2$) of a binomial distribution is $npq$

$\therefore \sigma^2 = npq = np(1 - p) = np - np^2 = f(p)$ say

For $f(p)$ to be maximum

$$f'(p) = 0 \text{ and } f''(p) < 0$$

Now $f'(p) = n - 2np = 0 \Rightarrow p = \frac{1}{2}$

Also $f''(p) = -2n < 0$

$\therefore f(p)$ is maximum at $p = \frac{1}{2}$

Thus maximum variance is at $p = \frac{1}{2}$, $q = \frac{1}{2}$

i.e. maximum variance $= npq = n.\frac{1}{2}.\frac{1}{2} = \frac{n}{4}$ $\therefore$ Variance $\leq \frac{n}{4}$

**Example 2** 6 dice are thrown 729 times. How many times would you expect at least three dice to show 1 or 2?

**Solution**: Here the Binomial Distribution (B.D.) is given by $N(q + p)^n$

Where $p = \frac{2}{6} = \frac{1}{3}$, $q = \frac{2}{3}$, $n = 6$, $N = 729$

$\therefore$ B.D. is given by $729 \left(\frac{2}{3} + \frac{1}{3}\right)^6$ and if $X$ is random variable showing number of successes, then

$$P(X \geq 3) = 729 \left[ {}^6C_3 \left(\frac{2}{3}\right)^3 \left(\frac{1}{3}\right)^3 + {}^6C_4 \left(\frac{2}{3}\right)^2 \left(\frac{1}{3}\right)^4 + {}^6C_5 \left(\frac{2}{3}\right)^1 \left(\frac{1}{3}\right)^5 + {}^6C_6 \left(\frac{1}{3}\right)^6 \right]$$

$$= \frac{729}{3^6} [160 + 60 + 12 + 1] = 233$$

**Example 3** A die is tossed 3 times. Find mean and variance of number of successes if getting 5 or 6 is considered as success.

**Solution**: Here $p = \frac{2}{6} = \frac{1}{3}$, $q = \frac{2}{3}$, $n = 3$

$\therefore$ Mean $= np = \frac{3}{3} = 1$, Variance $= npq = 3 \times \frac{1}{3} \times \frac{2}{3} = \frac{2}{3}$

**Example 4** If the sum of mean and variance of Binomial Distribution is 4.8 for 5 trials. Find the distribution.

**Solution**: Given $np + npq = 4.8$, $n = 5$

$\Rightarrow np(1 + q) = 4.8$

$\Rightarrow 5(1 - q)(1 + q) = 4.8$ $\because n = 5$

$\Rightarrow 1 - q^2 = 9.6 \Rightarrow q^2 = \frac{1}{25} \Rightarrow q = \frac{1}{5}$

$\therefore q = \frac{1}{5}$, $p = \frac{4}{5}$ and distribution is given by $\left(\frac{1}{5} + \frac{4}{5}\right)^5$

**Example 5** For a binomial distribution; mean is 4 and standard deviation is $\sqrt{2}$. Find the distribution.

**Solution**: Given $np = 4$, $npq = 2$

$\Rightarrow \frac{npq}{np} = \frac{2}{4} \Rightarrow q = \frac{2}{4} = \frac{1}{2} \therefore p = \frac{1}{2}$

Also $np = 4$ or $\frac{n}{2} = 4 \Rightarrow n = 8$

$\therefore$ The distribution is given by $\left(\frac{1}{2} + \frac{1}{2}\right)^8$

**Example 6** Find the expected number of the defective bulbs in a lot of 100 bulbs; if one out of 5 bulbs is defective. Also find the standard deviation, coefficient of skewness $\gamma_1$ and determine whether the distribution curve is leptokurtic, mesokurtic or platykurtic.

**Solution:** We have $p = \frac{1}{5} = 0.2$, $q = 1 - 0.2 = 0.8$, $n = 1000$

Expected number of defective bulbs $= np = 100 \times 0.2 = 20$

Also standard deviation $= \sqrt{npq} = \sqrt{100 \times 0.2 \times 0.8} = 4$

$\gamma_1 = \frac{(q-p)}{\sqrt{npq}} = \frac{(0.8-0.2)}{4} = 0.15$

$\beta_2 = 3 + \frac{1-6pq}{npq} = 3 + \frac{1-6(0.2)(0.8)}{16} = 3.0025$

$\therefore$ The curve is a bit leptokurtic.

**Example 7** If the probability of success of an event is $\frac{1}{20}$; how many trials are required in order that the probability of getting at least one success, is just greater than $\frac{1}{2}$ ?

**Solution:** Here $p = \frac{1}{20}$, $q = \frac{19}{20}$

Let $n$ be the required number of trials such that the probability of getting at least one success, is just greater than $\frac{1}{2}$

i.e. $P(X \geq 1) > \frac{1}{2}$

$\Rightarrow 1 - P(X = 0) > \frac{1}{2}$

$\Rightarrow P(X = 0) < 1 - \frac{1}{2}$

$\Rightarrow {}^nC_0 \left(\frac{19}{20}\right)^n \left(\frac{1}{20}\right)^0 < \frac{1}{2}$

$\Rightarrow \left(\frac{19}{20}\right)^n < \frac{1}{2}$

$\Rightarrow n \log_{10} \frac{19}{20} < \log_{10} \frac{1}{2}$

$\Rightarrow n > \frac{\log_{10} \frac{1}{2}}{\log_{10} \frac{19}{20}} \quad \because \log_{10} \frac{19}{20} < 0$

$\Rightarrow n > \frac{-0.3010}{-0.02228} = 13.5099 \quad \therefore n = 14$

**Example 8** The probability of a man hitting a target is 1/3. How many times must he take the shot so that the probability of hitting the target at least once is less than 90%?

**Solution**: Here $p = \frac{1}{3}, q = \frac{2}{3}$, Let $n$ be the number of shots so that the probability of hitting the target, at least once, is less than 90%.

i.e. $P(X \geq 1) < \frac{9}{10}$

$\Rightarrow 1 - P(X = 0) < \frac{9}{10}$

$\Rightarrow P(X = 0) > 1 - \frac{9}{10}$

$\Rightarrow {}^nC_0 \left(\frac{2}{3}\right)^n \left(\frac{1}{3}\right)^0 > \frac{1}{10}$

$\Rightarrow \left(\frac{2}{3}\right)^n > \frac{1}{10}$

$\Rightarrow n \log_{10}\frac{2}{3} > \log_{10}\frac{1}{10}$

$\Rightarrow n < \frac{\log_{10}\frac{1}{10}}{\log_{10}\frac{2}{3}}$ $\quad \because \log_{10}\frac{2}{3} < 0$

$\Rightarrow n < \frac{-1}{-0.1761} = 5.6786 \quad \therefore n = 5$

**Example 9** Assuming that half the population are consumers of chocolates, so that the chances of an individual being consumer is $\frac{1}{2}$ and assuming that each of the 25 surveyors takes 10 individuals to see whether they are consumers. How many surveyors would you expect to report that three or less people were consumers?

**Solution**: The probability of an individual to be consumer $(p) = \frac{1}{2}, \therefore q = \frac{1}{2}$

Also $n = 10$, $N = 25$

$\therefore$ B.D. is given by $25\left(\frac{1}{2} + \frac{1}{2}\right)^{10}$ and if $X$ is random variable showing number of successes, then

$P(X \leq 3) = 25\left[{}^{10}C_0 \left(\frac{1}{2}\right)^{10} + {}^{10}C_1 \left(\frac{1}{2}\right)^9 \left(\frac{1}{2}\right)^1 + {}^{10}C_2 \left(\frac{1}{2}\right)^8 \left(\frac{1}{2}\right)^2 + {}^{10}C_3 \left(\frac{1}{2}\right)^7 \left(\frac{1}{2}\right)^3\right]$

$= 25\left(\frac{1}{2}\right)^{10}\left[1 + 10 + \frac{10 \times 9}{2!} + \frac{10 \times 9 \times 8}{3!}\right] = 25(0.171875) = 4.3$

So we can expect four surveyors to report that three or less people were consumers.

**Example 10**: Fit a binomial distribution to the following data and compare theoretical frequencies with actual ones

| x | 0 | 1 | 2 | 3 | 4 | 5 | 6 | 7 | 8 | 9 |
|---|---|---|---|---|---|---|---|---|---|---|
| f | 6 | 20 | 28 | 12 | 8 | 6 | 0 | 0 | 0 | 0 |

**Solution**: Mean of the given distribution = $\frac{\Sigma fx}{\Sigma f}, \Sigma f = 80$

$$= \frac{0+20+56+36+32+30+0}{80} = \frac{87}{40} = 2.175$$

Let mean of binomial distribution to be fitted $= np = 2.175$

Also $n = 10 \therefore p = 0.2175 \quad q = 1 - 0.2175 = 0.7825$

$\therefore$ B.D. is given by $80(0.7825 + 0.2175)^{10}$

Theoretical frequencies using binomial distribution are given in the table below:

| x | $P(r) = {}^nC_r\, q^{n-r} p^r$ | Theoretical frequencies $(f) = 80 \times P(r)$ |
|---|---|---|
| 0 | ${}^{10}C_0\,(0.7825)^{10}(0.2175)^0 = 0.086$ | $6.9 = 7\text{(say)}$ |
| 1 | ${}^{10}C_1\,(0.7825)^9(0.2175)^1 = 0.239$ | $19.1 = 19\text{(say)}$ |
| 2 | ${}^{10}C_2\,(0.7825)^8(0.2175)^2 = 0.299$ | $23.9 = 24\text{(say)}$ |
| 3 | ${}^{10}C_3\,(0.7825)^7(0.2175)^3 = 0.22$ | $17.8 = 18\text{(say)}$ |
| 4 | ${}^{10}C_4\,(0.7825)^6(0.2175)^4 = 0.11$ | $8.6 = 9\text{ (say)}$ |
| 5 | ${}^{10}C_5\,(0.7825)^5(0.2175)^5 = 0.04$ | $2.9 = 3\text{ (say)}$ |
| 6 | ${}^{10}C_6\,(0.7825)^4(0.2175)^6 = 0.008$ | $0.66 = 0\text{ (say)}$ |
| 7 | ${}^{10}C_7\,(0.7825)^3(0.2175)^7 = 0.001$ | $0.11 = 0\text{ (say)}$ |
| 8 | ${}^{10}C_8\,(0.7825)^2(0.2175)^8 = 0$ | 0 |
| 9 | ${}^{10}C_9\,(0.7825)^1(0.2175)^9 = 0$ | 0 |

## 3.3 Poisson Distribution

**Result**: Poisson distribution with $P(r) = \frac{e^{-\lambda}\lambda^r}{r!}$ is a limiting case of Binomial distribution, under the conditions **i.** $n \to \infty$ **ii.** $p \to 0$ **iii.** $np = \lambda$ is finite

**Proof**: In a Binomial distribution

$$P(r) = {}^nC_r\, q^{n-r} p^r$$

$$= {}^nC_r\,(1-p)^{n-r} p^r$$

$$= {}^nC_r\,\left(1-\frac{\lambda}{n}\right)^{n-r}\left(\frac{\lambda}{n}\right)^r \qquad \because np = \lambda$$

$$= \frac{n(n-1)(n-2)\cdots(n-(r-1))}{r!}\left(1-\frac{\lambda}{n}\right)^{n-r}\left(\frac{\lambda}{n}\right)^r$$

$$= \frac{1\left(1-\frac{1}{n}\right)\left(1-\frac{2}{n}\right)\cdots\left(1-\frac{(r-1)}{n}\right)}{\left(1-\frac{\lambda}{n}\right)^r r!}\left(1-\frac{\lambda}{n}\right)^n \lambda^r$$

Taking limit as $n \to \infty$

$$P(r) = \frac{\lambda^r}{r!}\lim_{n\to\infty}\left(1-\frac{\lambda}{n}\right)^n$$

$$= \frac{\lambda^r}{r!}\lim_{n\to\infty}\left[\left(1-\frac{\lambda}{n}\right)^{\frac{-n}{\lambda}}\right]^{-\lambda}$$

$$\therefore P(r) = \frac{e^{-\lambda}\lambda^r}{r!} \qquad \because \lim_{n\to\infty}\left(1+\frac{1}{x}\right)^x = e$$

### 3.3.1 Mean of Poisson Distribution

$$\text{Mean} = \sum_{r=0}^{\infty} rP(r) = \sum_{r=0}^{\infty} \frac{re^{-\lambda}\lambda^r}{r!}$$

$$= \frac{e^{-\lambda}\lambda^1}{1!} + \frac{2e^{-\lambda}\lambda^2}{2!} + \frac{3e^{-\lambda}\lambda^3}{3!} + \frac{4e^{-\lambda}\lambda^4}{4!} + \cdots$$

$$= e^{-\lambda}\lambda\left(1 + \lambda + \frac{\lambda^2}{2!} + \frac{\lambda^3}{3!} + \cdots\right)$$

$$= e^{-\lambda}\lambda e^{\lambda} = \lambda$$

### 3.3.2 Variance of Poisson Distribution

$$\text{Variance} = \sum_{r=0}^{n} r^2 P(r) - (\text{mean})^2$$

Now $\sum_{r=0}^{n} r^2 P(r) = \sum_{r=0}^{\infty} \frac{r^2 e^{-\lambda}\lambda^r}{r!}$

$$= \frac{e^{-\lambda}\lambda^1}{1!} + \frac{2^2 e^{-\lambda}\lambda^2}{2!} + \frac{3^2 e^{-\lambda}\lambda^3}{3!} + \frac{4^2 e^{-\lambda}\lambda^4}{4!} + \cdots$$

$$= e^{-\lambda}\lambda\left(1 + 2\lambda + \frac{3\lambda^2}{2!} + \frac{4\lambda^3}{3!} + \cdots\right)$$

$$= e^{-\lambda}\lambda\left[\left(1 + \lambda + \frac{\lambda^2}{2!} + \frac{\lambda^3}{3!} + \cdots\right) + \left(\lambda + \lambda^2 + \frac{3\lambda^3}{3!} + \cdots\right)\right]$$

$$= e^{-\lambda}\lambda\left[\left(1 + \lambda + \frac{\lambda^2}{2!} + \frac{\lambda^3}{3!} + \cdots\right) + \lambda\left(1 + \lambda + \frac{\lambda^2}{2!} + \frac{\lambda^3}{3!} + \cdots\right)\right]$$

$$= e^{-\lambda}\lambda[e^{\lambda} + \lambda e^{\lambda}]$$

$$\Rightarrow \sum_{r=0}^{n} r^2 P(r) = e^{-\lambda}\lambda e^{\lambda}[1 + \lambda] = \lambda + \lambda^2$$

∴ Variance $= \lambda + \lambda^2 - \lambda^2 = \lambda$

### 3.3.3 Moments of Poisson Distribution

Since Poisson distribution is a limiting case of binomial distribution, therefore mean and moments may be obtained from Binomial distribution by taking $np = \lambda$, $p \to 0$ and $q \to 1$ as limit $n \to \infty$

Mean $= \lim_{n \to \infty} np = \lambda$,

$$\mu_1 = 0$$

$$\mu_2 = \sigma^2 = \lim_{n \to \infty} npq = \lambda \quad \because \mu_2 = \text{variance } (\sigma^2)$$

Similarly $\mu_3 = \lim_{n \to \infty} npq(q - p) = \lambda.1(1 - 0) = \lambda$

$$\mu_4 = \lim_{n \to \infty} npq[1 + 3pq(n - 2)]$$

$$= \lim_{n \to \infty} npq[1 + 3npq - 6pq]$$

$$= \lambda.1[1 + 3\lambda.1 - 6.0.1]$$

$$= \lambda + 3\lambda^2$$

$$\beta_1 = \frac{u_3^2}{u_2^3} = \frac{\lambda^2}{\lambda^3} = \frac{1}{\lambda}, \quad \beta_2 = \frac{\mu_4}{u_2^2} = \frac{\lambda + 3\lambda^2}{\lambda^2} = \frac{1}{\lambda} + 3$$

**Example11**: If the standard deviation of a Poisson variate $X$ is $\sqrt{3}$, then find the probability that $X$ is strictly positive.

**Solution**: Here variance $(\lambda) = 3$

$$P(X = r) = \frac{e^{-\lambda}\lambda^r}{r!} = \frac{e^{-3}3^r}{r!}, \quad r = 0, 1, 2, 3, \cdots$$

The probability that $X$ is strictly positive is:

$$P(X > 0) = 1 - P(X = 0)$$
$$= 1 - \frac{e^{-3}3^0}{0!} = 1 - e^{-3}$$

**Example12**: *i.* If the probability of a bad reaction from a certain injection is 0.001, determine the chance that out of 1000 individuals more than two will have bad reaction.

*ii.* A manufacturer who produces medicine bottles finds that 0.1% of the bottles are defective. The bottles are packed in boxes containing 500 bottles. Find the probability that in 100 such boxes how many are expected to contain (*a*) no defective bottle (*b*) at least two defective bottles.

**Solution**: *i.* Here $p = 0.001$, $n = 1000$  $\lambda = np = 1000 \times 0.001 = 1$

$$P(X = r) = \frac{e^{-\lambda}\lambda^r}{r!} = \frac{e^{-1}(1)^r}{r!}$$

Probability that more than two individuals will have bad reaction is given by:

$$P(X > 2) = 1 - P(X \leq 2) = \{1 - [P(X = 0) + P(X = 1) + P(X = 2)]\}$$
$$= \left\{1 - e^{-1}\left[\frac{(1)^0}{0!} + \frac{(1)^1}{1!} + \frac{(1)^2}{2!}\right]\right\} = 1 - e^{-1}\left[\frac{5}{2}\right] = 1 - 0.9197 = 0.0803$$

*ii.* Here $p = 0.1\% = \frac{0.1}{100} = 0.001$, $n = 500$  $\lambda = np = 500 \times 0.001 = 0.5$

$$P(X = r) = \frac{e^{-\lambda}\lambda^r}{r!} = \frac{e^{-0.5}(0.5)^r}{r!}$$

(*a*) Probability of zero defective bottles in a box of 500 bottles is given by:

$$P(X = 0) = \frac{e^{-0.5}(0.5)^0}{0!} = e^{-0.5} = 0.6065$$

∴ Number of boxes having no defective bottle out of 100 boxes
$$= 100 \times 0.6065 = 60.65 \text{ approx}$$

(*b*) Probability of at least 2 defective bottles in a box of 500 is given by:

$$P(X \geq 2) = 1 - P(X < 2) = \{1 - [P(X = 0) + P(X = 1)]\}$$
$$= \left\{1 - e^{-0.5}\left[\frac{(0.5)^0}{0!} + \frac{(0.5)^1}{1!}\right]\right\} = 1 - e^{-0.5}[1.5] = 0.0902$$

∴ Number of boxes having at least 2 defective bottles out of 100 boxes
$$= 100 \times 0.0902 = 9.02 \text{ approx}$$

**Example13** In a certain factory producing tyres, there is a small chance of 1 in 500 tyres to be defective. The tyres are supplied in lots of 10. Using Poisson distribution, calculate the approximate number of lots containing   *i.* no defective

ii. at least one defective tyre in a consignment of 10,000 lots.

**Solution**: Here $p = \frac{1}{500}$, $n = 10$ $\lambda = np = \frac{10}{500} = 0.02$

$$P(X = r) = \frac{e^{-\lambda}\lambda^r}{r!} = \frac{e^{-0.02}(0.02)^r}{r!}$$

i. Probability of no defective tyre in a lot is given by:

$$P(X = 0) = \frac{e^{-0.02}(0.02)^0}{0!} = e^{-0.02} = 0.9802$$

∴ Number of lots containing no defective tyre = $10000 \times 0.9802 = 9802$ approx

ii. Probability of at least one defective tyre in a lot is given by:

$$1 - P(X = 0) = 1 - 0.9802 = 0.0198$$

∴ Number of lots containing at least one defective tyre

$= 10000 \times 0.0198 = 198$ Approx

**Example14**: A skilled typist kept a record of his mistakes made per day during 300 working days. Fit a Poisson distribution to compare theoretical frequencies with actual ones

| Mistakes per day | 0 | 1 | 2 | 3 | 4 | 5 | 6 |
|---|---|---|---|---|---|---|---|
| Number of days | 143 | 90 | 42 | 12 | 9 | 3 | 1 |

**Solution**: Mean of the given distribution = $\frac{\Sigma fx}{\Sigma f}$, $\Sigma f = 300$

$$= \frac{0+90+84+36+36+15+6}{300} = \frac{89}{100} = 0.89 = \lambda$$

| Mistakes per day | $P(r) = \frac{e^{-\lambda}\lambda^r}{r!}$ | Theoretical frequency $300 \times P(r)$ |
|---|---|---|
| 0 | $\frac{e^{-(0.89)}(0.89)^0}{0!} = 0.411$ | 123.3=123 (say) |
| 1 | $\frac{e^{-(0.89)}(0.89)^1}{1!} = 0.365$ | 109.5=110 (say) |
| 2 | $\frac{e^{-(0.89)}(0.89)^2}{2!} = 0.163$ | 48.9=49 (say) |
| 3 | $\frac{e^{-(0.89)}(0.89)^3}{3!} = 0.048$ | 14.4=14 (say) |
| 4 | $\frac{e^{-(0.89)}(0.89)^4}{4!} = 0.011$ | 3.3=3 (say) |
| 5 | $\frac{e^{-(0.89)}(0.89)^5}{5!} = 0.002$ | 0.6=1 (say) |
| 6 | $\frac{e^{-(0.89)}(0.89)^6}{6!} = 0.0003$ | 0.09=0 (say) |

**Example15**: The distribution of number of road accidents per day in a city is Poisson with mean 5. Find the number of days in a year when there will be

i. at most 2 accidents  ii. between 3 and 5 accidents

**Solution**: Here $\lambda = np = 5$

$$P(X = r) = \frac{e^{-\lambda}\lambda^r}{r!} = \frac{e^{-5}(5)^r}{r!}$$

i. Probability of at most 2 accidents per day is given by:

$$P(X \leq 2) = P(X = 0) + P(X = 1) + P(X = 2)$$

$$= e^{-5}\left[\frac{(5)^0}{0!} + \frac{(5)^1}{1!} + \frac{(5)^2}{2!}\right] = 0.1247$$

∴ Number of days in a year having at most 2 accidents per day

$$= 365 \times 0.1247 = 45.5 \text{ approx}$$

ii. Probability of 3 to 5 accidents per day is given by:

$$P(3 \leq X \leq 5) = P(X = 3) + P(X = 4) + P(X = 5)$$

$$= e^{-5}\left[\frac{(5)^3}{3!} + \frac{(5)^4}{4!} + \frac{(5)^5}{5!}\right] = 0.4913$$

∴ Number of days in a year having 3 to 5 accidents per day

$$= 365 \times 0.4913 = 179.3 \text{ approx}$$

## 3.4 Normal Distribution

The normal distribution developed by Gauss is a continuous distribution and is very useful in practical applications. It can be considered as the limiting form of the Binomial Distribution when the number of trials $(n)$, is very large and neither $p$ nor $q$ is very small. The probability curve of a normal variate $x$ with mean $\mu$ and standard deviation $\sigma$ is given by:

$$p(x) = \frac{1}{\sigma\sqrt{2\pi}} e^{-\frac{1}{2}\left(\frac{x-\mu}{\sigma}\right)^2}, -\infty < x < \infty$$

Any normal variate $x$ with mean $\mu$ and standard deviation $\sigma$ is changed to a standard normal variate $z = \frac{x-\mu}{\sigma}$, and hence the probability density function of $z$ is given by: $\phi(z) = \frac{1}{\sqrt{2\pi}} e^{-\frac{1}{2}z^2}$, $-\infty < z < \infty$

The normal distribution with mean $\mu$ and variance $\sigma^2$ is denoted by $N(\mu, \sigma^2)$.

Adjoining figure shows a normal distribution curve for standard normal variate $z$.

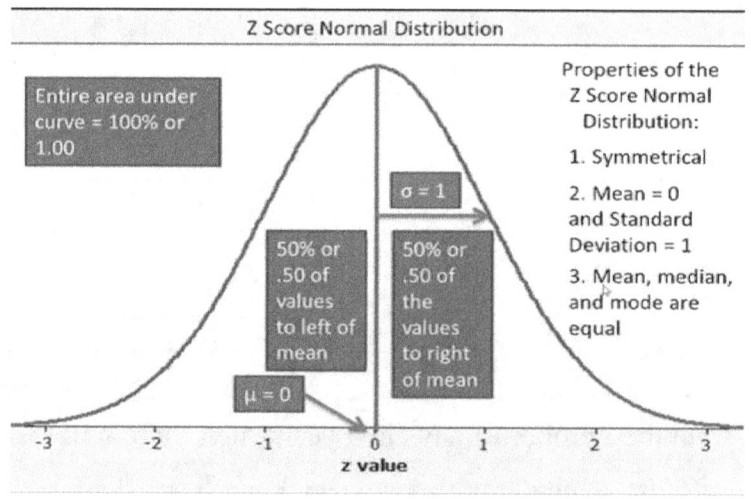

**Properties of Normal Curve**:

- The graph of $p(x)$ or $\phi(z)$ is a bell shaped curve.
- Since the distribution is symmetrical, mean, mode and median coincide at $x = \mu$ or $z = 0$. Also $\beta_1 = 0 \Rightarrow \gamma_1 = 0$ and $\beta_2 = 3 \Rightarrow \gamma_2 = 0$

- The ordinate at $x = \mu$ or $z = 0$, divides the whole area into two equal parts. Also since the total area under the probability curve is 1, area to the right of the ordinate as well as to the left of the ordinate at $x = \mu$ or $z = 0$ is 0.5.
- Area under the curve between the ordinates $a < z < b$ gives the probability of variate $z$ taking values between $a$ and $b$. Area is concentrated more towards the middle and goes on decreasing on the either sides of the curve, i.e. tails, but never becomes zero. The curve never intersects $x$-axis at any finite point. i.e. $x$-axis is its asymptote.
- Since the curve is symmetrical about mean. The first quartile $Q_1$ and the third quartile $Q_3$ lie at the same distance on the two sides of the mean $\mu$. The distance of any quartile from $\mu$ is $0.6745\,\sigma$ units. Thus

  $Q_1 = \mu - 0.6745\sigma$ or $\mu - \frac{2}{3}\sigma$

  $Q_3 = \mu + 0.6745\sigma$ or $\mu + \frac{2}{3}\sigma$

- Area under the normal curve is distributed as follows:

  Area between $x = \mu - \sigma$ and $x = \mu + \sigma$ is 18.27%

  Area between $x = \mu - 2\sigma$ and $x = \mu + 2\sigma$ is 95.45%

  Area between $x = \mu - 3\sigma$ and $x = \mu + 3\sigma$ is 99.73%

- Points of inflection are $\mu \pm \sigma$
- Quartile deviation $(QD)$ is

  $\frac{1}{2}(Q_3 - Q_1)\ \frac{2}{3}\sigma$

- Mean deviation $(MD)$ is $\frac{4}{5}\sigma$

  $\therefore QD:MD:SD \equiv 10:12:15$

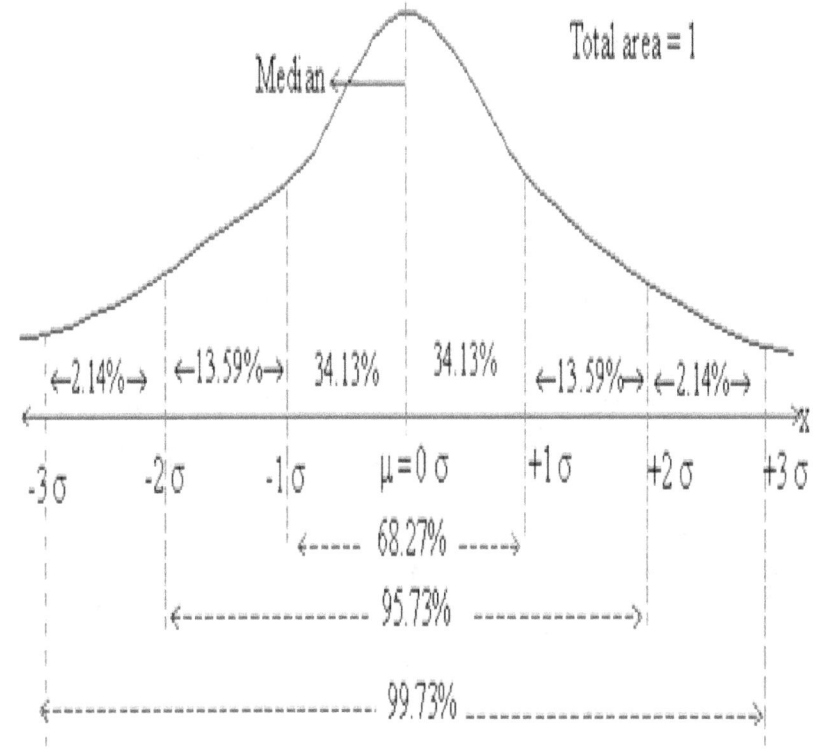

- Since the distribution is symmetrical, all the moments of odd order about mean are zero, i.e. $\mu_{2n+1} = 0$, $n = 1, 2, 3, \cdots$
- Moments of even order are given by:

  $\mu_{2n} = 1.3.5 \ldots (2n-1)\sigma^{2n}$, $n = 1, 2, 3, \cdots$

Putting $n = 1$ and 2; $\mu_2 = \sigma^2$, $\mu_4 = 3\sigma^4$, $\beta_1 = \frac{u_3^2}{u_2^3} = 0$, $\beta_2 = \frac{\mu_4}{u_2^2} = \frac{3\sigma^4}{\sigma^4} = 3$

**Using Normal Distribution tables:** We first convert the variate $x$ into standard normal variate $z$; using the relation $z = \frac{x-\mu}{\sigma}$ and find the limits of $z$-score corresponding to the given limits of the variate $x$. Normal Distribution tables are broadly of two types; either of the two may be used to determine the area (probability) for the standard normal variate $z$ usually given in the range $(-3.49 \leq z \leq 3.49)$.

The first type of table gives the area covered by standard normal variate $z$ between the ordinates 0 to $z$ as shown in adjoining figure. This table covers more than 0.499 units of area on positive side of the curve on the whole. If value of z is negative, we may use the symmetrical property of the normal curve, i.e area in the region $z < -a$ is same as area $z > a$ where $a$ is any positive number within the range $(0, 3.49)$.

The second table gives the area covered by standard normal variate $z$ from $-\infty$ to $z$ as shown in the given figure. This table covers more than 0.998 units of area on the whole.

### 3.4.1 Moments of Normal Distribution

**Result**: Odd order moments about mean are zero

**Proof**: $r^{th}$ moment about the mean $\bar{x}$ of a continuous distribution is given by:

$$\mu_r = \int_{-\infty}^{\infty} \frac{f(x-\bar{x})^r}{n} dx$$

∴ Odd order moments of normal distribution with mean $\mu$ are given by:

$$\mu_{2n+1} = \int_{-\infty}^{\infty} (x-\mu)^{2n+1} p(x) dx \quad \because \frac{f}{n} = p(x)$$

$$= \frac{1}{\sigma\sqrt{2\pi}} \int_{-\infty}^{\infty} (x-\mu)^{2n+1} e^{-\frac{1}{2}\left(\frac{x-\mu}{\sigma}\right)^2} dx \quad \because p(x) = \frac{1}{\sigma\sqrt{2\pi}} e^{-\frac{1}{2}\left(\frac{x-\mu}{\sigma}\right)^2}$$

$$= \frac{\sigma^{2n+1}}{\sqrt{2\pi}} \int_{-\infty}^{\infty} z^{2n+1} e^{-\frac{1}{2}z^2} dz \text{ Putting } \frac{x-\mu}{\sigma} = z, \ dx = \sigma \, dz$$

$$= 0, \ z^{2n+1} e^{-\frac{1}{2}z^2} \text{ being an odd function of } z.$$

**Result**: Even order moments about mean are given by:

$$\mu_{2n} = 1.3.5 \ldots (2n-1)\sigma^{2n}, \ n = 1, 2, 3, \cdots$$

**Proof**: $\mu_{2n} = \int_{-\infty}^{\infty} (x-\mu)^{2n} p(x) dx$

$$= \frac{1}{\sigma\sqrt{2\pi}} \int_{-\infty}^{\infty} (x-\mu)^{2n} e^{-\frac{1}{2}\left(\frac{x-\mu}{\sigma}\right)^2} dx \quad \because p(x) = \frac{1}{\sigma\sqrt{2\pi}} e^{-\frac{1}{2}\left(\frac{x-\mu}{\sigma}\right)^2}$$

$$= \frac{\sigma^{2n}}{\sqrt{2\pi}} \int_{-\infty}^{\infty} z^{2n} e^{-\frac{1}{2}z^2} dz \text{ Putting } \frac{x-\mu}{\sigma} = z \Rightarrow dx = \sigma \, dz$$

$$= \frac{2\sigma^{2n}}{\sqrt{2\pi}} \int_{0}^{\infty} z^{2n} e^{-\frac{1}{2}z^2} dz, \ z^{2n} e^{-\frac{1}{2}z^2} \text{ being even function of } z.$$

$$= \frac{2\sigma^{2n}}{\sqrt{2\pi}} \int_{0}^{\infty} (2t)^n e^{-t} (2t)^{-\frac{1}{2}} dt \text{ Putting } \frac{z^2}{2} = t \Rightarrow zdz = dt$$

$$= \frac{2^n \sigma^{2n}}{\sqrt{\pi}} \int_{0}^{\infty} e^{-t} t^{n-\frac{1}{2}} dt$$

$$\mu_{2n} = \frac{2^n \sigma^{2n}}{\sqrt{\pi}} \Gamma\left(n+\frac{1}{2}\right) \quad \because \Gamma n = \int_{0}^{\infty} e^{-t} t^{n-1} dt$$

Again changing $n$ to $(n-1)$

$$\mu_{2n-2} = \frac{2^{n-1} \sigma^{2n-2}}{\sqrt{\pi}} \Gamma\left(n-\frac{1}{2}\right)$$

$$\therefore \frac{\mu_{2n}}{\mu_{2n-2}} = 2\sigma^2 \frac{\Gamma\left(n+\frac{1}{2}\right)}{\Gamma\left(n-\frac{1}{2}\right)} = 2\sigma^2 \frac{\left(n-\frac{1}{2}\right)\Gamma\left(n-\frac{1}{2}\right)}{\Gamma\left(n-\frac{1}{2}\right)} = 2\sigma^2 \left(n-\frac{1}{2}\right)$$

$$\therefore \mu_{2n} = \sigma^2(2n-1)\mu_{2n-2}$$

$$\Rightarrow \mu_{2n} = \sigma^2(2n-1)\sigma^2(2n-3)\mu_{2n-4}$$
$$= \sigma^2(2n-1)\sigma^2(2n-3)\sigma^2(2n-5)\mu_{2n-6}$$
$$\vdots$$
$$\therefore \mu_{2n} = 1.3.5 \ldots (2n-5)(2n-3)(2n-1)\sigma^{2n}, n = 1, 2, 3, \cdots$$

**Example 16** For a normal distribution the mean 20 and the standard deviation 15, find *i.* $Q_1$ and $Q_3$ *ii.* Mean deviation *iii.* the inter quartile range.

**Solution:** *i.* For a normal distribution
$$Q_1 = \mu - \frac{2}{3}\sigma = 20 - \frac{2}{3}(15) = 10$$
$$Q_3 = \mu + \frac{2}{3}\sigma = 20 + \frac{2}{3}(15) = 30$$

*ii.* Mean deviation is $\frac{4}{5}\sigma = \frac{4}{5}(15) = 12$

*iii.* The inter quartile range is $Q_3 - Q_1 = 20$

**Example 17** Find $p$, mean and the standard deviation of the normal distribution given by $y = pe^{-\left(\frac{x^2}{8} - x + 2\right)}$

**Solution:** Rewriting $\left(\frac{x^2}{8} - x + 2\right)$ as $\frac{1}{2}\left(\frac{x^2 - 8x + 16}{4}\right) = \frac{1}{2}\left(\frac{x-4}{2}\right)^2$

$$\therefore y = pe^{-\frac{1}{2}\left(\frac{x-4}{2}\right)^2}$$

Comparing with $y = \frac{1}{\sigma\sqrt{2\pi}} e^{-\frac{1}{2}\left(\frac{x-\mu}{\sigma}\right)^2}$, we get $\mu = 4$, $\sigma = 2$, $p = \frac{1}{\sigma\sqrt{2\pi}} = \frac{1}{2\sqrt{2\pi}}$

**Example 18** If $X$ is a random variable with mean 20 and standard deviation 5, find the probabilities that (i) $15 \le X \le 25$ (ii) $|X - 20| > 5$

**Solution:** Given that $X$ is a random variable with mean $\mu = 20$ and S.D. $\sigma = 5$.

i.e. $X \sim N(100, 25)$ and $z = \frac{X-\mu}{\sigma} = \frac{X-20}{5} \Rightarrow X = 20 + 5z$

(i) $P(15 \le X \le 25)$

$= P(15 < 20 + 5z < 25)$

$= P\left(\frac{15-20}{5} < z < \frac{25-20}{5}\right)$

$= P(-1 < z < 1) = 2P(0 < z < 1)$

$= 2(0.3413)$ Using normal table

$= 0.6826$

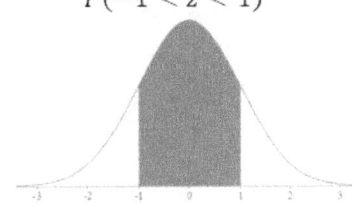

$P(-1 < z < 1)$

(ii) $P(|X - 20| > 5) = 1 - P(|X - 20| \le 5)$

Now $P(|X - 20| \le 5) = P(-5 \le X - 20 \le 5) = P(15 \le X \le 25)$

$= 0.6826$ using part (i)

$\therefore P(|X - 20| > 5) = 1 - 0.6826 = 0.3174$

**Example19** Fifteen hundred candidates appeared in a certain examination, having maximum marks as 100. It was found that the marks are normally distributed with mean 55 and standard deviation as 10.5. Determine approximately the number of candidates who were passed with distinction, i.e. 75% and above marks.

**Solution:** Let the random variable $X$ denote the marks obtained out of 100.

Then $X$ is a random variable with mean $\mu = 55$ and S.D. $\sigma = 10.5$.

i.e. $X \sim N(55, 10.5^2)$ and $z = \frac{X-\mu}{\sigma} = \frac{X-55}{10.5}$

$\Rightarrow X = 55 + 10.5z$

$P(X \geq 75)$

$= P(55 + 10.5z \geq 75)$

$= P\left(z \geq \frac{75-55}{10.5}\right) = P(z \geq 1.9)$

$= 0.5 - P(0 < z < 1.9)$

$= 0.5 - 0.4713 = 0.0287$

$P(z \geq 1.9)$

∴ Number of candidates who passed with distinction is:

$1500 \times 0.0287 = 43.05$ i.e. 43 approximately

**Example 20** The daily wages of 1000 workers are normally distributed with mean 100\$ and standard deviation 5\$. Estimate the number of workers whose daily wages will be:

(i) between 100\$ and 105\$      (ii) between 96\$ and 105\$

(iii) more than 110\$      (iv) more than 110\$

(v) Also estimate the daily wages of 100 highest paid workers.

**Solution:** Let the random variable $X$ denote the daily wages of workers in dollars.

Then $X$ is a random variable with mean $\mu = 100$ and S.D. $\sigma = 5$.

i.e. $X \sim N(100, 25)$ and $z = \frac{X-\mu}{\sigma} = \frac{X-100}{5}$

$\Rightarrow X = 100 + 5z$

(i) $P(100 < X < 105)$

$= P(100 < 100 + 5z < 105)$

$= P\left(\frac{100-100}{5} < z < \frac{105-100}{5}\right)$

$= P(0 < z < 1) = 0.3413$ using Z table

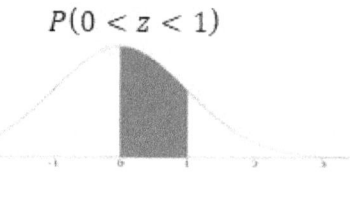

$P(0 < z < 1)$

(ii) $P(96 < X < 105)$

$= P(96 < 100 + 5z < 105)$

$= P\left(\frac{96-100}{5} < z < \frac{105-100}{5}\right)$

$= P(-0.8 < z < 1)$

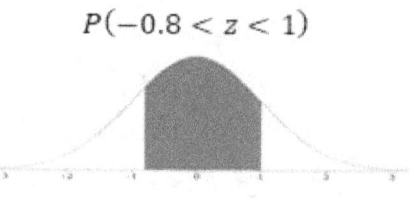

$P(-0.8 < z < 1)$

$$= P(0 < z < 0.8) + P(0 < z < 1)$$
$$= 0.2881 + 0.3413 = 0.6294 \text{ using } Z \text{ table}$$

(iii) $P(X > 110)$
$$= P(100 + 5z > 110)$$
$$= P\left(z > \frac{110-100}{5}\right)$$
$$= P(z > 2)$$
$$= 0.5 - P(0 < z < 2)$$
$$= 0.5 - 0.4772 = 0.0228 \text{ using } Z \text{ table}$$

(iv) $P(X < 92)$
$$= P(100 + 5z < 92)$$
$$= P\left(z < \frac{92-100}{5}\right)$$
$$= P(z < -1.6)$$
$$= P(z > 1.6) \text{ (Symmetry)}$$
$$= 0.5 - P(0 < z < 1.6)$$
$$= 0.5 - 0.4452 = 0.0548 \text{ using } Z \text{ table}$$

(v) Proportion of 100 highest paid workers is $\frac{100}{1000} = \frac{1}{10} = 0.1$

To determine $X = r$ such that $P(X > r) = 0.1$

When $X = r$, $z = \frac{r-100}{5} = z_1$ (say)

$\therefore P(z > z_1) = 0.1$

$\Rightarrow P(0 < z < z_1) = 0.5 - 0.1 = 0.4$

From normal distribution table, $z_1 = 1.28$ approx.

$\therefore z = \frac{r-100}{5} = 1.28$

$\Rightarrow r = 100 + 5 \times 1.28 = 106.4$

Hence the lowest daily wages of 100 highest paid workers are 106.4\$

**Example 21** If in a normal distribution 7% of the items are under 35 and 89% are under 63. What are the mean and standard deviations of the distribution?

**Solution**: Let $X$ be normally distributed with mean $\mu$ and standard deviation $\sigma$ such that $z = \frac{X-\mu}{\sigma}$.
Also given that $P(X < 35) = 0.07$ and $P(X < 63) = 0.89$
Let $z_1$ be a standard normal variate corresponding to $X = 35$,

$\therefore z_1 = \frac{35-\mu}{\sigma}$ and $P(z_1 < 35) = 0.07$

$\Rightarrow P(-z_1 > 35) = 0.07$

$\Rightarrow P(0 < -z_1 < z) = 0.5 - 0.07 = 0.43$

$\Rightarrow -z_1 = 1.48$ From normal distribution table

or $z_1 = \frac{35-\mu}{\sigma} = -1.48$

$\Rightarrow \mu - 1.48\sigma = 35$ ...①

Again let $z_2$ be a standard normal variate corresponding to $X = 63$

$\therefore z_2 = \frac{63-\mu}{\sigma}$ and $P(z_2 < 63) = 0.89$

$\Rightarrow P(0 < z_2 < z) = 0.89 - 0.5 = 0.39$

$\Rightarrow z_2 = \frac{63-\mu}{\sigma} = 1.23$ from normal distribution table

$\Rightarrow \mu + 1.23\sigma = 63$ ...②

Solving ① and ②, we get: $\mu = 109.6, \sigma = 50.4$

**Example 21**: Fit a normal curve to the following distribution

| x | 2 | 4 | 6 | 8 | 10 |
|---|---|---|---|---|----|
| f | 1 | 4 | 6 | 4 | 1  |

**Solution**: Let $x$ be normally distributed with mean $\mu$ and standard deviation $\sigma$ such that $z = \frac{X-\mu}{\sigma}$.

$\mu = \frac{\Sigma fx}{\Sigma f} = \frac{2+16+36+32+10}{16} = 6$

Also $\sigma = \sqrt{\frac{\Sigma fx^2}{\Sigma f} - \left(\frac{\Sigma fx}{\Sigma f}\right)^2} = \sqrt{\frac{(4+64+216+256+100)}{16} - 36} = \sqrt{40-36} = 2$

Again $x$ has to be a continuous variable to follow normal distribution, therefore taking $x$ as mid value of an interval $(x_1, x_2)$

| x | $(x_1, x_2)$ | $(z_1, z_2)$ | $P(z) \equiv$ Area under the curve in the interval $(z_1, z_2)$ | Theoretical frequency $16 \times P(z)$ |
|---|---|---|---|---|
| 2 | (1, 3) | (-2.5, -1.5) | 0.0606 | 0.9696 |
| 4 | (3, 5) | (-1.5, -0.5) | 0.2417 | 3.8672 |
| 6 | (5, 7) | (-0.5, 0.5) | 0.3829 | 6.1264 |
| 8 | (7, 9) | (0.5, 1.5) | 0.2417 | 3.8672 |
| 10 | (9, 11) | (1.5, 2.5) | 0.0606 | 0.9696 |

### Exercise 3

1. In 256 sets of 12 tosses of a coin, in how many cases can one expect 8 heads and 4 tails?
2. In a precision bombing attack, there are 50% chances that any one bomb will hit the target. If two direct hits are required to destroy the target completely, how many bombs must be dropped to provide a 99% or more chances to completely destroy the target.

3. Comment on the statement: For a binomial distribution mean is 5 and standard deviation is 3.
4. If the probability of a defective item is 0.02, find the probability that at most 5 defective items will be found out in a box of 200 items.
5. Six coins are tossed 6400 times. Using Poisson distribution, find the approximate probability of getting six heads 2 times.
6. In a certain factory making a machine part, probability of it being defective is 0.002. If the part is supplied in packs of 10, use Poisson distribution to calculate the approximate number of packets containing no defective and one defective machine part in a consignment of 10,000 packets.
7. A car hire firm has 2 cars, which are hired on daily basis. The number of demands for a car on each day follows Poisson distribution with mean 1.5. Calculate the proportion of days on which neither car is used and the proportion of days on which some demand is refused, given that $e^{-1.5} = 0.2231$.
8. If in a Poisson distribution $P(X = 0) = P(X = 1) = a$, show that $a = \frac{1}{e}$.
9. If $X$ is a random variable with mean 30 and standard deviation 5, find the probabilities that (i) $X > 45$ (ii) $|X - 30| \geq 5$
10. If in a normal distribution 31% of the items are under 45 and 8% are over 64. What are the mean and standard deviation of the distribution?
11. In an intelligence test given to 1000 children, the average score is 42 with the standard deviation 24. Find i. the number of children whose score exceeds 60

ii. the number of children whose score lie between 20 and 40.

12. Assuming mean height of the soldiers to be 68.22 inches with a variance of 10.8 square inches, how many soldiers in the regiment of 1000 would you expect to be over 6 feet tall, given that area under the standard normal curve between $z = 0$ to $z = 1.15$ is 0.3746.
13. Fit a binomial distribution for the given data:

| x | 0 | 1 | 2 | 3 | 4 | 5 | 6 | 7 | 8 |
|---|---|---|---|---|---|---|---|---|---|
| f | 0 | 5 | 9 | 22 | 25 | 26 | 14 | 4 | 1 |

## Answers

1. 31
2. 11
3. Variance cannot be greater than mean in case of a binomial distribution.
4. 0.7845
5. $5000\, e^{-100}$
6. 9802, 196
7. 0.2231, 0.1913
9. 0.0014, 0.3174
10. 50, 10
11. 227, 287
12. 125
13. 0.4, 3.3, 11.6, 23.2, 29, 23.2, 11.6, 3.3, 0.4

*Chapter 4: Curve Fitting & Correlation*

# Curve Fitting & Correlation

## 4.1 Introduction

The process of constructing an approximate curve $y = f(x)$, which fit best to a given discrete set of points $(x_i, y_i)$, $i = 1, 2, 3, \ldots, n$ is called curve fitting. Curve fitting and interpolation are closely associated procedures. In interpolation, the fitted function should pass through all given data points; whereas curve fitting methodologically fits a unique curve to the data points, which may or may not lie on the fitted curve. The difference between interpolation and curve fitting; while attempting to fit a linear function; is illustrated in the adjoining figure.

## 4.2 Principle of Least Squares

The principle of least squares is one of the most popular methods for finding the curve of best fit to a given data set $(x_i, y_i)$, $i = 1, 2, 3, \ldots, n$.

Let $y = f(x)$ be the equation of the curve to be fitted to the given set of points $P_1(x_1, y_1)$, $P_2(x_2, y_2), \cdots, P_n(x_n, y_n)$.

Then at a point $x = x_i$, the observed value of the ordinate is $P_i M_i = y_i$ say and let the expected (theoretical) value be $f(x_i)$, shown by $L_i M_i$ in the adjoining figure.

The difference between the observed and expected values is the error $e_i = P_i L_i$

Then $e_1 = y_1 - f(x_1)$

$e_2 = y_2 - f(x_2)$

$\vdots$

$e_n = y_n - f(x_n)$

Squaring each error $e_i$ (to take care of negative errors) and adding, we get

$E = e_1^2 + e_2^2 + \cdots + e_n^2 = \sum_{i=1}^{n} e_i^2$

$= \sum_{i=1}^{n}(y_i - f(x_i))^2$. The curve of best fit is that for which $E$ is minimum. This is called the Principle of least squares.

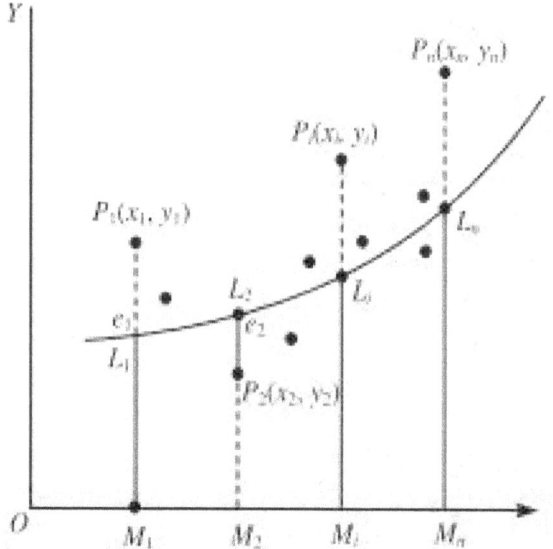

## 4.2.1 Fitting a straight line

Let $y = ax + b$ be the straight line to be fitted to the given set of data points $(x_1, y_1)$, $(x_2, y_2), ..., (x_n, y_n)$.

Then $e_i = y_i - f(x_i) = y_i - (ax_i + b)$, $i = 1, 2, ..., n$

$\therefore E = \sum_{i=1}^{n} e_i^2 = \sum_{i=1}^{n}(y_i - (ax_i + b))^2$

Now by the principle of least square, for the curve of best fit, $E$ is minimum

$\therefore \frac{\partial E}{\partial a} = 0$ and $\frac{\partial E}{\partial b} = 0$

$\therefore \frac{\partial E}{\partial a} = 0$ and $\frac{\partial E}{\partial b} = 0$

$\Rightarrow \sum_{i=1}^{n} 2(y_i - (ax_i + b))(-x_i) = 0$ $\qquad \Rightarrow \sum_{i=1}^{n} 2(y_i - ax_i - b)(-1) = 0$

$\Rightarrow \sum_{i=1}^{n}(y_i x_i - ax_i^2 - bx_i) = 0$ $\qquad \Rightarrow \sum_{i=1}^{n}(y_i - ax_i - b) = 0$

$\Rightarrow \sum_{i=1}^{n} y_i x_i = a\sum_{i=1}^{n} x_i^2 + b\sum_{i=1}^{n} x_i$ $\qquad \Rightarrow \sum_{i=1}^{n} y_i = a\sum_{i=1}^{n} x_i + nb$
... ① ... ②

Since $x_i, y_i$ are known set of values, equation ① and ② are two equation with variables $a$ and $b$, ignoring the suffices, ② and ① can be rewritten as

$\sum y = a\sum x + nb$ ... ③

and $\sum xy = a\sum x^2 + b\sum x$ ... ④

③ and ④ are known as Normal equation for fitting a straight line $y = ax + b$

> If the equation of line is taken as $y = a + bx$
> we get normal equations as: $\sum y = na + b\sum x$ and $\sum xy = a\sum x + b\sum x^2$

**Example 1** By the method of least squares, find a straight line that best fits the following data points.

| $x$ | 0 | 1 | 2 | 3 | 4 |
|---|---|---|---|---|---|
| $y$ | 1.0 | 2.9 | 4.8 | 6.7 | 8.6 |

**Solution**: Let line of best fit be given by $y = ax + b$ ... ①

Normal equations are given by:

$\sum y = a\sum x + nb$ ... ②

and $\sum xy = a\sum x^2 + b\sum x$ ... ③

Calculating $\sum x, \sum y, \sum xy$ and $\sum x^2$

| $x$ | $y$ | $xy$ | $x^2$ |
|---|---|---|---|
| 0 | 1.0 | 0 | 0 |
| 1 | 2.9 | 2.9 | 1 |
| 2 | 4.8 | 9.6 | 4 |
| 3 | 6.7 | 20.1 | 9 |
| 4 | 8.6 | 34.4 | 16 |
| $\sum x = 10$ | $\sum y = 24$ | $\sum xy = 67.0$ | $\sum x^2 = 30$ |

Substituting values of $\sum x, \sum y, \sum xy$ and $\sum x^2$ in ② and ③

$$\Rightarrow 24 = 10a + 5b \qquad \ldots ④$$

and $\qquad 67 = 30a + 10b \qquad \ldots ⑤$

Solving ④ and ⑤, we get $a = 1.9$ and $b = 1$

Substituting in ①, line of best fit is $y = 1.9x + 1$

**Example 2** Fit a straight line to following data

| $x$ | 0 | 1 | 2 | 3 | 4 |
|---|---|---|---|---|---|
| $y$ | 1.0 | 1.8 | 3.3 | 4.5 | 6.3 |

**Solution**: Let line of best fit be given by $y = ax + b \qquad \ldots ①$

Normal equations are given by:

$$\sum y = a\sum x + nb \qquad \ldots ②$$

and $\qquad \sum xy = a\sum x^2 + b\sum x \qquad \ldots ③$

Calculating $\sum x, \sum y, \sum xy$ and $\sum x^2$

| $x$ | $y$ | $xy$ | $x^2$ |
|---|---|---|---|
| 0 | 1.0 | 0 | 0 |
| 1 | 1.8 | 1.8 | 1 |
| 2 | 3.3 | 6.6 | 4 |
| 3 | 4.5 | 13.5 | 9 |
| 4 | 6.3 | 25.2 | 16 |
| $\sum x = 10$ | $\sum y = 16.9$ | $\sum xy = 47.1$ | $\sum x^2 = 30$ |

Substituting values of $\sum x, \sum y, \sum xy$ and $\sum x^2$ in ② and ③

$$\Rightarrow 16.9 = 10a + 5b \qquad \ldots ④$$

and $\qquad 47.1 = 30a + 10b \qquad \ldots ⑤$

Solving ④ and ⑤, we get $a = \frac{133}{100} = 1.33$ and $b = \frac{18}{25} = 0.72$

Substituting in ①, line of best fit is $y = 1.33x + 0.72$

**Example 3** If $F$ is the force required to lift a load $W$, by means of a pulley, fit a linear expression $F = a + bW$ against the following data:

| W | 50 | 70 | 100 | 120 |
|---|----|----|-----|-----|
| F | 12 | 15 | 21  | 25  |

**Solution**: Line for best fit is given as $F = a + bW$ ... ①

Normal equations are given by:

$$\sum F = na + b\sum W \qquad \ldots ②$$

and $$\sum WF = a\sum W + b\sum W^2 \qquad \ldots ③$$

| W | F | WF | W² |
|---|---|----|----|
| 50 | 12 | 600 | 2500 |
| 70 | 15 | 1050 | 4900 |
| 100 | 21 | 2100 | 10000 |
| 120 | 25 | 3000 | 14400 |
| $\sum W = 340$ | $\sum F = 73$ | $\sum WF = 6750$ | $\sum W^2 = 31800$ |

Substituting values of $\sum W, \sum F, \sum WF$ and $\sum W^2$ in ② and ③

$$\Rightarrow 73 = 4a + 340b \qquad \ldots ④$$

and $$6750 = 340a + 31800b \qquad \ldots ⑤$$

Solving ④ and ⑤, we get $a = 2.2759$ and $b = 0.1879$

Substituting in ①, line of best fit is $F = 2.2759 + 0.1879W$

### 4.2.2 Fitting a parabola

Let $y = ax^2 + bx + c$ be the parabola to be fitted to the given set of data points $(x_1, y_1), (x_2, y_2), \ldots, (x_n, y_n)$.

Then $e_i = y_i - f(x_i) = y_i - (ax_i^2 + bx_i + c)$, $i = 1, 2, \ldots, n$

$$\therefore E = \sum_{i=1}^{n} e_i^2 = \sum_{i=1}^{n}(y_i - (ax_i^2 + bx_i + c))^2$$

Now by the principle of least square, for the curve of best fit, $E$ is minimum

$$\therefore \frac{\partial E}{\partial a} = 0, \quad \frac{\partial E}{\partial b} = 0 \text{ and } \frac{\partial E}{\partial c} = 0$$

Solving we get normal equations as:

$$\sum y = a\sum x^2 + b\sum x + nc$$

$$\sum xy = a\sum x^3 + b\sum x^2 + c\sum x$$

$$\sum x^2 y = a\sum x^4 + b\sum x^3 + c\sum x^2$$

**Example 4** Fit a parabola $y = ax^2 + bx + c$ to the given data

| x | 10 | 12 | 15 | 23 | 20 |
|---|----|----|----|----|----|
| y | 14 | 17 | 23 | 25 | 21 |

**Solution**: Let the parabola of best fit be given by $y = ax^2 + bx + c$ ... ①

Normal equations are given by:

$$\sum y = a\sum x^2 + b\sum x + nc \qquad ...②$$

$$\sum xy = a\sum x^3 + b\sum x^2 + c\sum x \qquad ...③$$

$$\sum x^2 y = a\sum x^4 + b\sum x^3 + c\sum x^2 \qquad ...④$$

| x | y | xy | $x^2$ | $x^2 y$ | $x^3$ | $x^4$ |
|---|---|----|----|---------|-------|-------|
| 10 | 14 | 140 | 100 | 1400 | 1000 | 10000 |
| 12 | 17 | 204 | 144 | 2448 | 1728 | 20736 |
| 15 | 23 | 345 | 225 | 5175 | 3375 | 50625 |
| 23 | 25 | 575 | 529 | 13225 | 12167 | 279841 |
| 20 | 21 | 420 | 400 | 8400 | 8000 | 160000 |
| $\sum x = 80$ | $\sum y = 100$ | $\sum xy = 1684$ | $\sum x^2 = 1398$ | $\sum x^2 y = 30684$ | $\sum x^3 = 26270$ | $\sum x^4 = 521202$ |

Substituting values of $\sum x, \sum y, \sum xy$ and $\sum x^2$ in ② and ③ and ④

$$\Rightarrow 100 = 1398a + 80b + 5c \qquad ...⑤$$

$$1684 = 26270a + 1398b + 80c \qquad ...⑥$$

$$30648 = 521202a + 26270b + 1398c \qquad ...⑦$$

Solving ⑤ ⑥ and ⑦, we get $a = -0.07$, $b = 3.01$, $c = -8.73$

Substituting in ①, parabola of best fit is $y = -0.07 x^2 + 3.01x - 8.73$

**Example 5** Fit a 2$^{nd}$ parabola to the given data

| x | 1 | 3 | 4 | 6 | 8 | 9 | 11 | 14 |
|---|---|---|---|---|---|---|----|----|
| y | 1 | 2 | 4 | 4 | 5 | 7 | 8  | 9  |

**Solution**: Let the parabola of best fit be given by $y = ax^2 + bx + c$ ... ①

| x | y | xy | $x^2$ | $x^2 y$ | $x^3$ | $x^4$ |
|---|---|----|-------|---------|-------|-------|
| 1 | 1 | 1 | 1 | 1 | 1 | 1 |
| 3 | 2 | 6 | 9 | 18 | 27 | 81 |
| 4 | 4 | 16 | 16 | 64 | 64 | 256 |
| 6 | 4 | 24 | 36 | 144 | 216 | 1296 |
| 8 | 5 | 40 | 64 | 320 | 512 | 4096 |
| 9 | 7 | 63 | 81 | 567 | 729 | 6561 |
| 11 | 8 | 88 | 121 | 968 | 1331 | 14641 |
| 14 | 9 | 126 | 196 | 1764 | 2744 | 38416 |
| $\sum x = 56$ | $\sum y = 40$ | $\sum xy = 364$ | $\sum x^2 = 524$ | $\sum x^2 y = 3846$ | $\sum x^3 = 5624$ | $\sum x^4 = 65348$ |

Normal equations are given by:

$$\sum y = a\sum x^2 + b\sum x + nc \quad \ldots ②$$

$$\sum xy = a\sum x^3 + b\sum x^2 + c\sum x \quad \ldots ③$$

$$\sum x^2 y = a\sum x^4 + b\sum x^3 + c\sum x^2 \quad \ldots ④$$

Substituting values of $\sum x, \sum y, \sum xy$ and $\sum x^2$ in ② and ③ and ④

$$\Rightarrow 40 = 524a + 56b + 8c \quad \ldots ⑤$$

$$364 = 5624a + 524b + 56c \quad \ldots ⑥$$

$$3846 = 65348a + 5624b + 524c \quad \ldots ⑦$$

Solving ⑤ ⑥ and ⑦, we get

$$a = \frac{103}{11229} = 0.009, \; b = \frac{8672}{11229} = 0.77, \; c = \frac{4375}{22458} = 0.195$$

Substituting in ①, parabola of best fit is $y = 0.009\,x^2 + 0.77x + 0.195$

### 4.2.3 Change of Scale

If the data values are equispaced (with height ($h$)) and quite large for computation, simplification may be done by origin shifting as given below:

- When number of observations ($n$) is odd, take the origin at middle value of the table; say ($x_0$) and substitute $u = \frac{x - x_0}{h}$
- $y$ values if small; may be left unchanged; or we can shift them at average value of $y$ data $v = \frac{y - y_0}{h}$
- When number of observations ($n$) is even, take the origin as mean of two middle values, with new height $\frac{h}{2}$ and substitute $u = \frac{x - x_0}{h/2}$

**Example 6** The weight of a calf taken at end of every month is given below. Fit a straight line using the method of least squares. Also compute monthly growth rate.

| x | 1 | 2 | 3 | 4 | 5 | 6 | 7 | 8 | 9 | 10 |
|---|---|---|---|---|---|---|---|---|---|---|
| y | 52.5 | 58.7 | 65.0 | 70.2 | 75.4 | 81.1 | 87.2 | 95.5 | 102.2 | 108.4 |

**Solution:** Here $n = 10$ is even, $\therefore$ taking origin at $\frac{5+6}{2} = 5.5$ and new height as

$$\frac{h}{2} = 0.5 \; \therefore u = \frac{x - 5.5}{0.5} \text{ and let } v = y$$

Let line of best fit $y = ax + b$ be transformed to $v = Au + B \quad \ldots ①$

Normal equations are given by $\sum v = a\sum u + nb \quad \ldots ②$

and $\sum uv = a\sum u^2 + b\sum u \quad \ldots ③$

Calculating $\sum u, \sum u^2, \sum v$ and $\sum uv$

| $x$ | $u = \dfrac{x-5.5}{0.5}$ | $v = y$ | $uv$ | $u^2$ |
|---|---|---|---|---|
| 1 | -9 | 52.5 | -472.5 | 81 |
| 2 | -7 | 58.7 | -410.9 | 49 |
| 3 | -5 | 65.0 | -325.0 | 25 |
| 4 | -3 | 70.2 | -210.6 | 9 |
| 5 | -1 | 75.4 | -75.4 | 1 |
| 6 | 1 | 81.1 | 81.1 | 1 |
| 7 | 3 | 87.2 | 261.6 | 9 |
| 8 | 5 | 95.5 | 477.5 | 25 |
| 9 | 7 | 102.2 | 715.4 | 49 |
| 10 | 9 | 108.4 | 975.6 | 81 |
| | $\sum u = 0$ | $\sum v = 796.2$ | $\sum uv = 1016.8$ | $\sum u^2 = 330$ |

Substituting values of $\sum u, \sum v, \sum uv$ and $\sum u^2$ in ② and ③

$\Rightarrow 796.2 = 10B$ and $1016.8 = 330a$

$\therefore A = 3.081$ and $B = 79.62$

Substituting in ①, line of best fit is $v = 3.081u + 79.62$

$\Rightarrow y = 3.081\left(\dfrac{x-5.5}{0.5}\right) + 79.62$

$\therefore$ Line of best fit is $y = 6.162x + 45.729$

Average growth rate per month is given by: $\dfrac{dy}{dx} = 6.162$

**Example 6** Fit a $2^{nd}$ degree parabola for the following data:

| $x$ | 1929 | 1930 | 1931 | 1932 | 1933 | 1934 | 1935 | 1936 | 1937 |
|---|---|---|---|---|---|---|---|---|---|
| $y$ | 352 | 356 | 357 | 358 | 360 | 361 | 361 | 360 | 359 |

**Solution:** Since number of observations is odd and $h = 1$,

taking $x_0 = 1933, y_0 = 357, u = x - 1933, v = y - 357$

The equation $y = ax^2 + bx + c$ is transformed to $v = Au^2 + Bu + C$ ... ①

Normal equations are

$$\sum v = A\sum u^2 + B\sum u + 9c \quad \text{...②}$$
$$\sum uv = A\sum u^3 + B\sum u^2 + c\sum u \quad \text{...③}$$
$$\sum u^2 v = A\sum u^4 + B\sum u^3 + c\sum u^2 \quad \text{...④}$$

Calculating $\sum u, \sum u^2, \sum u^3, \sum u^4, \sum v, \sum uv$ and $\sum u^2 v$

| x | u | y | v | uv | $u^2$ | $u^2v$ | $u^3$ | $u^4$ |
|---|---|---|---|---|---|---|---|---|
| 1929 | -4 | 352 | -5 | 20 | 16 | -80 | -64 | 256 |
| 1930 | -3 | 356 | -1 | 3 | 9 | -9 | -27 | 81 |
| 1931 | -2 | 357 | 0 | 0 | 4 | 0 | -8 | 16 |
| 1932 | -1 | 358 | 1 | -1 | 1 | 1 | -1 | 1 |
| 1933 | 0 | 360 | 3 | 0 | 0 | 0 | 0 | 0 |
| 1934 | 1 | 361 | 4 | 4 | 1 | 4 | 1 | 1 |
| 1935 | 2 | 361 | 4 | 8 | 4 | 16 | 8 | 16 |
| 1936 | 3 | 360 | 3 | 9 | 9 | 27 | 27 | 81 |
| 1937 | 4 | 359 | 2 | 8 | 16 | 32 | 64 | 256 |
| | $\sum u = 0$ | | $\sum v = 11$ | $\sum uv = 51$ | $\sum u^2 = 60$ | $\sum u^2v = -9$ | $\sum u^3 = 0$ | $\sum u^4 = 708$ |

Substituting $\sum u, \sum u^2, \sum u^3, \sum u^4, \sum v, \sum uv$ and $\sum u^2v$ in ② and ③ and ④

$$\Rightarrow 11 = 60A + 9C \qquad \ldots ⑤$$

$$51 = 60B \qquad \ldots ⑥$$

$$-9 = 708A + 60C \qquad \ldots ⑦$$

Solving ⑤ ⑥ and ⑦, we get $A = \frac{-247}{924}$ and $B = \frac{17}{20}$, $C = \frac{694}{231}$

Substituting in ①, parabola of best fit is $v = \frac{-247}{924}x^2 + \frac{17}{20}x + \frac{694}{231}$

$$\Rightarrow y - 357 = \frac{-247}{924}(x - 1933)^2 + \frac{17}{20}(x - 1933) + \frac{694}{231}$$

$$\Rightarrow y = -0.267x^2 + 1034.29x - 1000106.41$$

## 4.3 Correlation

Correlation is a measure of association between two variables; which may be dependent or independent. Whenever two variables $x$ and $y$ are so related; that increase in one is accompanied by an increase or decrease in the other, then the variables are said to be correlated. Coefficient of correlation ($r$) lies between $-1$ and $+1$, i.e. $-1 \leq r \leq 1$.

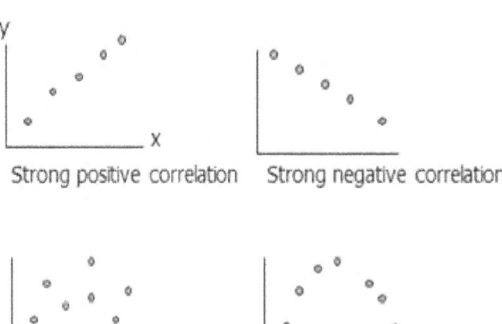

Strong positive correlation   Strong negative correlation

Zero correlation   Zero correlation

If $r$ is zero; no correlation between two variables, positive correlation ($0 < r \leq +1$); when both variables increase or decrease simultaneously, and negative correlation ($-1 \leq r < 0$); when increase in one is associated with decrease in other variable and vice-versa.

## 4.4 Karl Pearson Coefficient of Correlation

Coefficient of correlation ($r$) between two variables $x$ and $y$ is defined as

$$r = \frac{Covariance\ (x,y)}{\sqrt{Variance\ (x)}\sqrt{Variance\ (y)}} = \frac{\sum d_x d_y}{\sqrt{(\sum d_x^2)(\sum d_y^2)}} = \frac{\rho}{\sigma_x \sigma_y}$$

where $d_x = x - \bar{x}$, $d_y = y - \bar{y}$, $\bar{x}, \bar{y}$ are means of $x$ and $y$ data values.

$\rho = Cov(x,y) = \frac{\Sigma d_x d_y}{n}$ is the covariance between the variables $x$ and $y$.

Also $\sigma_x = \sqrt{\frac{\Sigma d_x^2}{n}}$ and $\sigma_y = \sqrt{\frac{\Sigma d_y^2}{n}}$

**Example 8** If $Cov(x,y) = 10$, $var(x) = 25$, $var(y) = 9$ find coefficient of correlation.

**Solution:** $r = \frac{Cov(x,y)}{\sqrt{Var(x)}\sqrt{Var(y)}} = \frac{10}{\sqrt{25}\sqrt{9}} = \frac{10}{5\times 3} = 0.67$

**Example 9** Calculate coefficient of correlation from the following data:

| x | 9 | 8 | 7 | 6 | 5 | 4 | 3 | 2 | 1 |
|---|---|---|---|---|---|---|---|---|---|
| y | 15 | 16 | 14 | 13 | 11 | 12 | 10 | 8 | 9 |

**Solution:** Karl Pearson coefficient of correlation $(r)$ is given by: $r = \frac{\Sigma d_x d_y}{\sqrt{(\Sigma d_x^2)(\Sigma d_y^2)}}$

where $d_x = x - \bar{x}$, $d_y = y - \bar{y}$, $\bar{x}, \bar{y}$ are means of $x$ and $y$ data values.

| $x$ | $d_x$ ($x-\bar{x}$) | $d_x^2$ | $y$ | $d_y$ ($y-\bar{y}$) | $d_y^2$ | $d_x d_y$ |
|---|---|---|---|---|---|---|
| 9 | 4 | 16 | 15 | 3 | 9 | 12 |
| 8 | 3 | 9 | 16 | 4 | 16 | 12 |
| 7 | 2 | 4 | 14 | 2 | 4 | 4 |
| 6 | 1 | 1 | 13 | 1 | 1 | 1 |
| 5 | 0 | 0 | 11 | -1 | 1 | 0 |
| 4 | -1 | 1 | 12 | 0 | 0 | 0 |
| 3 | -2 | 2 | 10 | -2 | 4 | 4 |
| 2 | -3 | 9 | 8 | -4 | 16 | 12 |
| 1 | -4 | 16 | 9 | -3 | 9 | 12 |
| $\Sigma x = 45$ $\bar{x} = 5$ | | $\Sigma d_x^2 = 60$ | $\Sigma y = 108$ $\bar{y} = 12$ | | $\Sigma d_y^2 = 60$ | $\Sigma d_x d_y = 57$ |

$\therefore r = \frac{\Sigma d_x d_y}{\sqrt{(\Sigma d_x^2)(\Sigma d_y^2)}} = \frac{57}{\sqrt{60\times 60}} = \frac{57}{60} = 0.95$

### 4.4.1 Shortcut Method for Karl Pearson Coefficient of Correlation

We can also find Karl Pearson Coefficient of Correlation by taking assumed means as shown:
If we take $d_x = x - a$; $d_y = y - b$, $a$ and $b$ are assumed means of $x$ and $y$ data values

Then $r = \dfrac{\Sigma d_x d_y - \frac{\Sigma d_x \Sigma d_y}{n}}{\sqrt{(\Sigma d_x^2) - \frac{1}{n}(\Sigma d_x)^2}\sqrt{(\Sigma d_y^2) - \frac{1}{n}(\Sigma d_y)^2}}$

- If $x_i$'s are equispaced with height $h$, we can take $d_x = \frac{x-a}{h}$
  Similarly if $y_i$'s are equispaced with height $k$, then $d_y = \frac{y-b}{k}$

**Example 10** Calculate coefficient of correlation from the following data:

| x | 1 | 3 | 5 | 7 | 8 | 10 |
|---|---|---|---|---|---|----|
| y | 8 | 12 | 15 | 17 | 18 | 20 |

**Solution**: Let $d_x = x - 7$, $d_y = y - 15$

$$r = \frac{\sum d_x d_y - \frac{\sum d_x \sum d_y}{n}}{\sqrt{(\sum d_x^2) - \frac{1}{n}(\sum d_x)^2} \sqrt{(\sum d_y^2) - \frac{1}{n}(\sum d_y)^2}}$$

Calculating $\sum d_x, \sum d_y, \sum d_x^2, \sum d_y^2$ and $\sum d_x d_y$

| x | $d_x$ (x − 7) | $d_x^2$ | y | $d_y$ (y − 15) | $d_y^2$ | $d_x d_y$ |
|---|---|---|---|---|---|---|
| 1 | -6 | 36 | 8 | -7 | 49 | 42 |
| 3 | -4 | 16 | 12 | -3 | 9 | 12 |
| 5 | -2 | 4 | 15 | 0 | 0 | 0 |
| 7 | 0 | 0 | 17 | 2 | 4 | 0 |
| 8 | 1 | 1 | 18 | 3 | 9 | 3 |
| 10 | 3 | 9 | 20 | 5 | 25 | 15 |
| | $\sum d_x = -8$ | $\sum d_x^2 = 66$ | | $\sum d_y = 0$ | $\sum d_y^2 = 96$ | $\sum d_x d_y = 72$ |

$$\therefore r = \frac{72 - \frac{(-8)(0)}{6}}{\sqrt{(66) - \frac{1}{6}(-8)^2} \sqrt{(96) - \frac{1}{6}(0)^2}} = 0.9879$$

### 4.4.2 Coefficient of Correlation of Bivariate Frequency Distribution

If given data is in the form of a bivariate frequency distribution,

Then $r = \dfrac{\sum f d_x d_y - \frac{\sum f d_x \sum f d_y}{n}}{\sqrt{(\sum f d_x^2) - \frac{1}{n}(\sum f d_x)^2} \sqrt{(\sum f d_y^2) - \frac{1}{n}(\sum f d_y)^2}}$, $n = \sum f$

**Example 11** Following table gives a bivariate distribution showing frequency of marks obtained according to age by a group of 52 students in an intelligent test:

| Marks \ Age | 16-18 | 18-20 | 20-22 | 22-24 |
|---|---|---|---|---|
| 10-20 | 2 | 1 | 1 | - |
| 20-30 | 3 | 2 | 3 | 2 |
| 30-40 | 3 | 4 | 5 | 6 |
| 40-50 | 2 | 2 | 3 | 4 |
| 50-60 | - | 1 | 2 | 2 |
| 60-70 | - | 1 | 2 | 1 |

Compute the correlation between marks and age of the students.

**Solution**: Let marks obtained by the students be denoted by $x$ and age by $y$, then coefficient of correlation ($r$) for the bivariate frequency distribution is given by:

$$r = \frac{\sum f d_x d_y - \frac{\sum f d_x \sum f d_y}{n}}{\sqrt{(\sum f d_x^2) - \frac{1}{n}(\sum f d_x)^2} \sqrt{(\sum f d_y^2) - \frac{1}{n}(\sum f d_y)^2}}, \quad n = \sum f$$

Let $d_x = \frac{x-a}{10}$, $d_y = \frac{y-b}{2}$, where $a$ & $b$ denote assumed mean classes

Here $a$ is taken as $30 - 40$, $b$ is taken as $18 - 20$

Also quantities in brackets denote $d_x d_y$ for each cell.

∴ $f d_x d_y$ for each cell is obtained by multiplying frequency of each cell with $d_x d_y$ and added across rows or columns to get $f d_x d_y$

| y \ x | 16-18 | 18-20 | 20-22 | 22-24 | F | $d_x$ | $f d_x$ | $f d_x^2$ | $f d_x d_y$ |
|---|---|---|---|---|---|---|---|---|---|
| 10-20 | 2 (2) | 1 (0) | 1 (-2) | - | 4 | -2 | -8 | 16 | 2 |
| 20-30 | 3 (1) | 2 (0) | 3 (-1) | 2 (-2) | 10 | -1 | -10 | 10 | -4 |
| 30-40 | 3 (0) | 4 (0) | 5 (0) | 6 (0) | 18 | 0 | 0 | 0 | 0 |
| 40-50 | 2 (-1) | 2 (0) | 3 (1) | 4 (2) | 11 | 1 | 11 | 11 | 9 |
| 50-60 | - | 1 (0) | 2 (2) | 2 (4) | 5 | 2 | 10 | 20 | 12 |
| 60-70 | - | 1 (0) | 2 (3) | 1 (6) | 4 | 3 | 12 | 36 | 12 |
| F | 10 | 11 | 16 | 15 | **52** | Totals | 15 | 93 | 31 |
| $d_y$ | -1 | 0 | 1 | 2 | Totals | | | | |
| $f d_y$ | -10 | 0 | 16 | 30 | 36 | | | | |
| $f d_y^2$ | 10 | 0 | 16 | 60 | 86 | | | | |
| $f d_x d_y$ | 5 | 0 | 8 | 18 | 31 | | | | |

$$\therefore r = \frac{31 - \frac{(15)(36)}{52}}{\sqrt{(93) - \frac{1}{52}(15)^2} \sqrt{(86) - \frac{1}{52}(36)^2}} = \frac{20.6154}{(9.4166)(7.8152)} = 0.2801$$

Thus there is a weak positive correlation between marks and age of the students.

4.4.3 Coefficient of Correlation by Rank differences
Rank correlation is used for attributes (like beauty, intelligence etc.) which cannot be measured quantitatively but can be provided with comparative ranks.

**Spearman's Rank Correlation** in given by: $r = 1 - \frac{6 \sum D^2}{n(n^2-1)}$, where $D = R_1 - R_2$

**Repeated or Tied Ranks**

If two or more observations in a data are equal, each observation is provided with an average rank and a correction factor is applied to correlation formula given as:

Correction Factor (C.F.) = $\sum m(m^2 - 1)$, $m$ is the number of times each observation is repeated.

**Spearman's Rank Correlation for repeated (tied) ranks is given by:**

$$r = 1 - \frac{6\left(\sum D^2 + \frac{1}{12}C.F.\right)}{n(n^2-1)}, \text{ where } D = R_1 - R_2$$

**Example12** Calculate the coefficient of correlation from the following data; given ranks of 10 students in English and Mathematics.

| Rank in English | 3 | 1 | 5 | 4 | 2 | 6 | 8 | 10 | 9 | 7 |
|---|---|---|---|---|---|---|---|---|---|---|
| Rank in Mathematics | 2 | 4 | 3 | 1 | 5 | 10 | 7 | 9 | 8 | 6 |

**Solution:** Since comparative ranks are given; instead of marks, using Spearman's Rank Correlation is given by: $r = 1 - \frac{6 \sum D^2}{n(n^2-1)}$, where $D = R_1 - R_2$

| Rank in English $R_1$ | Rank in Mathematics $R_2$ | $D = R_1 - R_2$ | $D^2$ |
|---|---|---|---|
| 3 | 2 | 1 | 1 |
| 1 | 4 | -3 | 9 |
| 5 | 3 | 2 | 4 |
| 4 | 1 | 3 | 9 |
| 2 | 5 | -3 | 9 |
| 6 | 10 | -4 | 16 |
| 8 | 7 | 1 | 1 |
| 10 | 9 | 1 | 1 |
| 9 | 8 | 1 | 1 |
| 7 | 6 | 1 | 1 |
| | | | $\sum D^2 = 52$ |

$$\therefore r = 1 - \frac{6(52)}{10(10^2-1)} = 0.6848$$

**Example13** Eight competitors in a beauty contest got marks (out of 10) by three judges as given below:

| Judge A | 9 | 6 | 5 | 10 | 3 | 1 | 4 | 2 |
|---|---|---|---|---|---|---|---|---|
| Judge B | 3 | 5 | 8 | 4 | 7 | 10 | 2 | 1 |
| Judge C | 6 | 4 | 9 | 8 | 1 | 2 | 3 | 10 |

Use rank correlation to discuss which pair of judges has the nearest approach to common tastes in beauty.

**Solution:** Since instead of ranks; marks are given by the three judges, converting the given data to comparative ranks for the eight competitors

| Judge A | | Judge B | | Judge C | | $D_{AB}$ | $D_{AB}^2$ | $D_{BC}$ | $D_{BC}^2$ | $D_{AC}$ | $D_{AC}^2$ |
|---|---|---|---|---|---|---|---|---|---|---|---|
| Marks | Rank | Marks | Rank | Marks | Rank | | | | | | |
| 9 | 2 | 3 | 6 | 6 | 4 | -4 | 16 | 2 | 4 | -2 | 4 |
| 6 | 3 | 5 | 4 | 4 | 5 | -1 | 1 | -1 | 1 | -2 | 4 |
| 5 | 4 | 8 | 2 | 9 | 2 | 2 | 4 | 0 | 0 | 2 | 4 |
| 10 | 1 | 4 | 5 | 8 | 3 | -4 | 16 | 2 | 4 | -2 | 4 |
| 3 | 6 | 7 | 3 | 1 | 8 | 3 | 9 | -5 | 25 | -2 | 4 |
| 1 | 8 | 10 | 1 | 2 | 7 | 7 | 49 | -6 | 36 | 1 | 1 |
| 4 | 5 | 2 | 7 | 3 | 6 | -2 | 4 | 1 | 1 | -1 | 1 |
| 2 | 7 | 1 | 8 | 10 | 1 | -1 | 1 | 7 | 49 | 6 | 36 |

Here $D_{AB}$ = Rank by Judge A − Rank by Judge B, also $\sum D_{AB}^2 = 100$

Similarly $D_{BC}$ = Rank by Judge B − Rank by Judge C, also $\sum D_{BC}^2 = 120$

$D_{AC}$ = Rank by Judge A − Rank by Judge C, also $\sum D_{AC}^2 = 58$

Rank Correlation between judges A and B is given by:

$$r_{AB} = 1 - \frac{6\sum D_{AB}^2}{n(n^2-1)} = 1 - \frac{6(100)}{8(8^2-1)} = -0.1905$$

Rank Correlation between judges B and C is given by:

$$r_{BC} = 1 - \frac{6\sum D_{BC}^2}{n(n^2-1)} = 1 - \frac{6(120)}{8(8^2-1)} = -0.4286$$

Rank Correlation between judges A and C is given by:

$$r_{AC} = 1 - \frac{6\sum D_{AC}^2}{n(n^2-1)} = 1 - \frac{6(58)}{8(8^2-1)} = 0.3095$$

Therefore Judges A and C have the nearest approach to common tastes in beauty, while Judges B and C have most different beauty tastes.

**Example 14:** Obtain rank correlation coefficient for following marks in economics ($x$) and Mathematics ($y$) out of 25 for eight students.

| $x$ | 20 | 24 | 12 | 20 | 10 | 12 | 24 | 20 |
|---|---|---|---|---|---|---|---|---|
| $y$ | 18 | 19 | 16 | 22 | 14 | 16 | 19 | 12 |

**Solution**: Converting data into ranks: Ranks of $x$ as $R_x$, Ranks of $y$ as $R_y$

| $x$ | $R_x$ | $y$ | $R_y$ | $D = R_x - R_y$ | $D^2$ |
|---|---|---|---|---|---|
| 20 | 4 | 18 | 4 | 0 | 0 |
| 24 | 1.5 | 19 | 2.5 | -1 | 1 |
| 12 | 6.5 | 16 | 5.5 | 1 | 1 |
| 20 | 4 | 22 | 1 | 3 | 9 |
| 10 | 8 | 14 | 7 | 1 | 1 |
| 12 | 6.5 | 16 | 5.5 | 1 | 1 |
| 24 | 1.5 | 19 | 2.5 | -1 | 1 |
| 20 | 4 | 12 | 8 | -4 | 16 |
| | | | | | $\Sigma D^2 = 30$ |

Correction Factor $= \Sigma m(m^2 - 1)$, $m$ is the number of times each data value is repeated

$\therefore$ C.F. $= 2(2^2 - 1) + 3(3^2 - 1) + 2(2^2 - 1) + 2(2^2 - 1) + 2(2^2 - 1) = 48$

Spearman's Rank Correlation for repeated ranks is given by:

$$r = 1 - \frac{6\left(\Sigma D^2 + \frac{1}{12}\text{C.F.}\right)}{n(n^2-1)}, \text{ where } D = R_x - R_y$$

$$\therefore r = 1 - \frac{6\left(30 + \frac{48}{12}\right)}{8(8^2-1)} = \frac{25}{42} = 0.595$$

**Example 15** Obtain rank correlation coefficient for following data

| $x$ | 68 | 64 | 75 | 50 | 64 | 80 | 75 | 40 | 55 | 64 |
|---|---|---|---|---|---|---|---|---|---|---|
| $y$ | 62 | 58 | 68 | 45 | 81 | 60 | 68 | 48 | 50 | 70 |

**Solution**: Converting data into ranks: Ranks of $x$ as $R_x$, Ranks of $y$ as $R_y$

| $X$ | $R_x$ | $Y$ | $R_y$ | $D = R_x - R_y$ | $D^2$ |
|---|---|---|---|---|---|
| 68 | 4 | 62 | 5 | -1 | 1 |
| 64 | 6 | 58 | 7 | -1 | 1 |
| 75 | 2.5 | 68 | 3.5 | -1 | 1 |
| 50 | 9 | 45 | 10 | -1 | 1 |
| 64 | 6 | 81 | 1 | 5 | 25 |
| 80 | 1 | 60 | 6 | -5 | 25 |
| 75 | 2.5 | 68 | 3.5 | -1 | 1 |
| 40 | 10 | 48 | 9 | 1 | 1 |
| 55 | 8 | 50 | 8 | 0 | 0 |
| 64 | 6 | 70 | 2 | 4 | 16 |
| | | | | | $\Sigma D^2 = 72$ |

Correction Factor (C.F.) $= \Sigma m(m^2 - 1)$, $m$ is the number of times each data value is repeated

$\therefore$ C.F. $= 2(2^2 - 1) + 3(3^2 - 1) + 2(2^2 - 1) = 36$

Spearman's Rank Correlation for repeated ranks is given by:

$$r = 1 - \frac{6\left(\Sigma D^2 + \frac{1}{12}\text{C.F.}\right)}{n(n^2-1)}, \text{ where } D = R_x - R_y \quad \therefore r = 1 - \frac{6\left(72 + \frac{36}{12}\right)}{10(10^2-1)} = \frac{6}{11} = 0.545$$

## 4.5 Linear Regression

Regression describes the functional relationship between dependent and independent variables; which helps us to make estimates of one variable from the other. Correlation quantifies the association between the two variables; whereas linear regression finds the best line that predicts $y$ from $x$ and also $x$ from $y$. The difference between correlation and regression is illustrated in the adjoining figure.

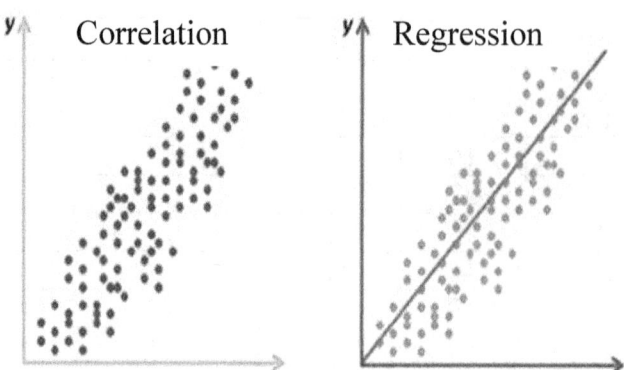

### 4.5.1 Lines of Regression:
If we plot the observations of the linear regression between two variables, actually two straight lines can approximately be drawn through the scatter diagram. One line estimates values of $y$ for specified values of $x$ (known as line of regression of $y$ on $x$); and other predicts values of $x$ from given values of $y$ (called line of regression of $x$ on $y$).

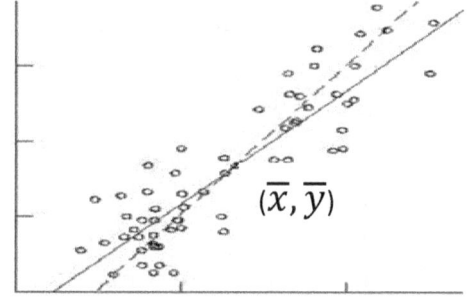

Let line of regression of $y$ on $x$ be represented by

$$y = a + bx \quad \ldots ①$$

Normal equations as derived by the method of least Square are:

$$\sum y = an + b\sum x \quad \ldots ②$$

and $\sum xy = a\sum x + b\sum x^2 \quad \ldots ③$

Dividing ② by $n$, we get

$$\frac{\sum y}{n} = a + b\frac{\sum x}{n} \Rightarrow \bar{y} = a + b\bar{x}$$

Where $\bar{x}$ and $\bar{y}$ are the means of $x$ series and $y$ series. This shows that $(\bar{x}, \bar{y})$ lies on the line of regression given by ①.

Again as $(\bar{x}, \bar{y})$ satisfies ①, shifting the origin to $(\bar{x}, \bar{y})$ in equation ③, we get

$$\sum(x - \bar{x})(y - \bar{y}) = a\sum(x - \bar{x}) + b\sum(x - \bar{x})^2$$

$$\Rightarrow \sum(x - \bar{x})(y - \bar{y}) = b\sum(x - \bar{x})^2 \qquad \because \sum(x - \bar{x}) = 0$$

$$\Rightarrow b = \frac{\sum(x-\bar{x})(y-\bar{y})}{\sum(x-\bar{x})^2} = \frac{\sum d_x d_y}{\sum d_x^2} \quad \ldots ④$$

Again $r = \frac{\sum d_x d_y}{\sqrt{(\sum d_x^2)(\sum d_y^2)}} = \frac{\sum d_x d_y}{n\sqrt{\frac{\sum d_x^2}{n}}\sqrt{\frac{\sum d_y^2}{n}}} = \frac{\sum d_x d_y}{n\sigma_x \sigma_y} \quad \because \sigma_x = \sqrt{\frac{\sum d_x^2}{n}}, \sigma_y = \sqrt{\frac{\sum d_y^2}{n}}$

Here $\sigma_x$, $\sigma_y$ are standard deviations of $x$ and $y$ data points respectively

$$\Rightarrow \sum d_x d_y = nr\sigma_x\sigma_y \qquad \ldots ⑤$$

Using ⑤ in ④, we get

$$b = \frac{n r \sigma_x \sigma_y}{\sum d_x^2} = \frac{r \sigma_x \sigma_y}{\sigma_x^2}$$

$$\Rightarrow b = \frac{r \sigma_y}{\sigma_x} \text{ which is slope of line of regression line of } y \text{ on } x$$

$$\therefore b_{yx} = \frac{r \sigma_y}{\sigma_x}, \ b_{yx} \text{ denotes slope of line of regression line of } y \text{ on } x.$$

Thus line of regression of $y$ on $x$ given by ①, passes through $(\bar{x}, \bar{y})$ and is having slope

$$b_{yx} = \frac{r \sigma_y}{\sigma_x}$$

∴ Equation of line of regression of $y$ on $x$ is given by $y - \bar{y} = b_{yx}(x - \bar{x})$

Similarly line of regression of $x$ on $y$ is given by: $x - \bar{x} = b_{xy}(y - \bar{y})$

where $b_{xy} = \frac{r \sigma_x}{\sigma_y}$ is slope of line of regression line of $x$ on $y$

Here $b_{xy}$ and $b_{yx}$ are known coefficients of regression and are connected by the relation:

$$b_{xy} b_{yx} = \left(\frac{r \sigma_x}{\sigma_y}\right)\left(\frac{r \sigma_y}{\sigma_x}\right) = r^2$$

### 4.5.1 Properties of Regression Coefficients

- As $\sqrt{b_{xy} b_{yx}} = r$, the coefficient of correlation is the geometric mean between the two regression coefficients.
- Since $\frac{b_{xy} + b_{yx}}{2} \geq \sqrt{b_{xy} b_{yx}} = r$, ∴ arithmetic mean of the two regression coefficients is greater than or equal to the correlation coefficient ($r$).
- If there is a perfect correlation between the two variables under consideration, then $b_{xy} = b_{yx} = r$; and the two lines of regression coincide. Converse is also true, i.e. if two lines of regression coincide, then there is a perfect correlation; $r = \pm 1$.
- Since $b_{xy} b_{yx} = r^2 > 0$, the signs of both regression coefficients $b_{xy}$ and $b_{yx}$ and coefficient of correlation ($r$) must be same; either all three negative or all positive.
- ∵ $b_{xy} b_{yx} = r^2 \leq 1$, if one of the regression coefficients is greater than unity, other must be less than unity.
- Point of intersection of two lines of regression is $(\bar{x}, \bar{y})$, Where $\bar{x}$ and $\bar{y}$ are the means of $x$ series and $y$ series.
- If both lines of regression cut each other at right angle, there is no correlation between the two variables; i.e. $r = 0$.

**Example 16** Prove that arithmetic mean of coefficients of regression is greater than the coefficient of correlation.

**Solution**: We know that $b_{xy} = \dfrac{r\sigma_x}{\sigma_y}$ and $b_{yx} = \dfrac{r\sigma_y}{\sigma_x}$

To prove $\dfrac{b_{xy}+b_{yx}}{2} > r$

or $\dfrac{1}{2}\left[\dfrac{r\sigma_x}{\sigma_y} + \dfrac{r\sigma_y}{\sigma_x}\right] > r$

or $\dfrac{1}{2}\left[\dfrac{\sigma_x^2 + \sigma_y^2}{\sigma_x\sigma_y}\right] > 1$

or $\left[\dfrac{\sigma_x^2 + \sigma_y^2}{\sigma_x\sigma_y}\right] - 2 > 0$

or $\sigma_x^2 + \sigma_y^2 - 2\sigma_x\sigma_y > 0$

or $[\sigma_x - \sigma_y]^2 > 0$

which is true

Note : A.M. $= r$ if $b_{xy} = b_{yx} = r = \pm 1$

## 4.5.2 Angle between the Lines of Regression

If $\theta$ be the acute angle between the two regression lines for two variables $x$ and $y$, then $\tan\theta = \dfrac{1-r^2}{r} \dfrac{\sigma_x \sigma_y}{\sigma_x^2 + \sigma_y^2}$

**Proof**: The two lines of regression are given by:

$y - \bar{y} = \dfrac{r\sigma_y}{\sigma_x}(x - \bar{x})$ ... ①

and $x - \bar{x} = \dfrac{r\sigma_x}{\sigma_y}(y - \bar{y})$ ... ②

If $m_1$ and $m_2$ are slopes of lines ① and ②, then

$\tan\theta = \dfrac{m_2 - m_1}{1 + m_1 m_2}$, where $m_1 = \dfrac{r\sigma_y}{\sigma_x}$, $m_2 = \dfrac{\sigma_y}{r\sigma_x}$

$\Rightarrow \tan\theta = \dfrac{\dfrac{\sigma_y}{r\sigma_x} - \dfrac{r\sigma_y}{\sigma_x}}{1 + \dfrac{r\sigma_y}{\sigma_x} \cdot \dfrac{\sigma_y}{r\sigma_x}} = \dfrac{\left(\dfrac{1}{r} - r\right)\dfrac{\sigma_y}{\sigma_x}}{1 + \dfrac{\sigma_y^2}{\sigma_x^2}} = \dfrac{1-r^2}{r} \dfrac{\sigma_x \sigma_y}{\sigma_x^2 + \sigma_y^2}$ ... ③

➢ When $r = 0$, $\tan\theta = \infty \Rightarrow \theta = \dfrac{\pi}{2}$ from ③

∴ when $r = 0$, the two lines of regression are perpendicular to each other.

➢ When $r = \pm 1$, $\tan\theta = 0 \Rightarrow \theta = 0$ from ③

∴ when $r = \pm 1$, the two lines of regression are coincident

**Example 17** Find the correlation coefficient between $x$ and $y$, when the two lines of regression are given by: $2x - 9y + 6 = 0$ and $x - 2y + 1 = 0$

**Solution:** Let the line of regression of $x$ on $y$ be $2x - 9y + 6 = 0$ ...①

Then the line of regression of $y$ on $x$ is $x - 2y + 1 = 0$ ...②

Now ① $\Rightarrow x = \frac{9}{2}y - 3$ $\therefore b_{xy} = \frac{9}{2}$

Also ② $\Rightarrow y = \frac{1}{2}x + \frac{1}{2}$ $\therefore b_{yx} = \frac{1}{2}$

$\therefore r = \sqrt{b_{xy} b_{yx}} = \sqrt{\frac{9}{2} \times \frac{1}{2}} = \frac{3}{2}$, which is not possible as $-1 \leq r \leq 1$

So our choice of regression lines is incorrect.

$\therefore$ Line of regression of $x$ on $y$ is $x - 2y + 1 = 0$

$\Rightarrow x = 2y - 1$ $\therefore b_{xy} = 2$

Also line of regression of $y$ on $x$ is $2x - 9y + 6 = 0$

$\Rightarrow y = \frac{2}{9}x + \frac{2}{3}$ $\therefore b_{yx} = \frac{2}{9}$

$\therefore r = \sqrt{b_{xy} b_{yx}} = \sqrt{2 \times \frac{2}{9}} = \frac{2}{3}$

**Example 18** The regression equations calculated from a given set of observations for two random variables are: $x = -0.4y + 6.4$ and $y = -0.6x + 4.6$

Calculate $\bar{x}, \bar{y}$ and $r$.

**Solution:** The two equations of regression are:

$x = -0.4y + 6.4$ ...①

$y = -0.6x + 4.6$ ...②

$\Rightarrow b_{xy} = -0.4$ and $b_{yx} = -0.6$

$\therefore r^2 = b_{xy} b_{yx} = 0.24$

$\Rightarrow r = \pm 0.49$

We know that the signs of $b_{xy}$, $b_{yx}$ and $r$ must be same

$\therefore r = -0.49$

Again we know that the point of intersection of two regression lines is $(\bar{x}, \bar{y})$

Therefore solving ① and ②, we get $\bar{x} = 6$, $\bar{y} = 1$

**Example 19** Find the regression line of $y$ on $x$ from the following data:

| $x$ | 1 | 3 | 4 | 6 | 8 | 9 | 11 | 14 |
|---|---|---|---|---|---|---|---|---|
| $y$ | 1 | 2 | 4 | 4 | 5 | 7 | 8 | 9 |

Also estimate the value of $y$, when $x = 10$

**Solution:** Let line of regression of $y$ on $x$ be:

$y = a + bx$ ...①

Then normal equations are given by:

$$\sum y = an + b \sum x \qquad \ldots ②$$
$$\text{and } \sum xy = a \sum x + b \sum x^2 \qquad \ldots ③$$

Calculating $\sum x, \sum y, \sum xy$ and $\sum x^2$

| $x$ | $y$ | $x^2$ | $xy$ |
|---|---|---|---|
| 1 | 1 | 1 | 1 |
| 3 | 2 | 9 | 6 |
| 4 | 4 | 16 | 16 |
| 6 | 4 | 36 | 24 |
| 8 | 5 | 64 | 40 |
| 9 | 7 | 81 | 63 |
| 11 | 8 | 121 | 88 |
| 14 | 9 | 196 | 126 |
| $\sum x = 56$ | $\sum y = 40$ | $\sum x^2 = 524$ | $\sum xy = 364$ |

Substituting values of $\sum x, \sum y, \sum xy$ and $\sum x^2$ in ② and ③

$$\Rightarrow 40 = 8a + 56b \qquad \ldots ④$$
$$\text{and } 364 = 56a + 524b \qquad \ldots ⑤$$

Solving ④ and ⑤, we get $a = \frac{6}{11}$ and $b = \frac{7}{11}$

Substituting in ①, line of regression of $y$ on $x$ is $y = \frac{6}{11} + \frac{7}{11}x$

$$\Rightarrow 7x - 11y + 6 = 0$$

Also at $x = 10$, $y = \frac{76}{11}$

**Example 20** From a partially destroyed lab data, following results were retrieved:

Lines of regression are:

$x = 0.45y + 5.35$ and $y = 0.8x + 6.6$, $\sigma_x^2 = 9$

Find $\bar{x}, \bar{y}, \sigma_y$ and $r$ for the existing data.

**Solution**: The two equations of regression are:

$$x = 0.45y + 5.35 \qquad \ldots ①$$
$$y = 0.8x + 6.6 \qquad \ldots ②$$

We know that the point of intersection of two regression lines is $(\bar{x}, \bar{y})$

Therefore solving ① and ②, we get $\bar{x} = 13$, $\bar{y} = 17$

Again ① $\Rightarrow b_{xy} = 0.45$ and $b_{yx} = 0.8$

$$\therefore r^2 = b_{xy} b_{yx} = 0.36$$

$\Rightarrow \quad r = \pm 0.6$

We know that the signs of $b_{xy}$, $b_{yx}$ and $r$ must be same

$\therefore r = 0.6$

Also $b_{yx} = \dfrac{r\,\sigma_y}{\sigma_x} \Rightarrow 0.8 = \dfrac{(0.6)\sigma_y}{3} \Rightarrow \sigma_y = \dfrac{0.8 \times 3}{0.6} = 4$

**Example 21** Following data depicts the statistical values of rainfall and production of wheat in a region for a specified time period.

|  | Mean | Standard Deviation |
|---|---|---|
| Production of Wheat (kg. per unit area) | 10 | 8 |
| Rainfall (cm) | 8 | 2 |

Estimate the production of wheat when rainfall is 9cm if correlation coefficient between production and rainfall is given to be 0.5.

**Solution**: Let the variables $x$ and $y$ denote production and rainfall respectively.

Given that $\bar{x} = 10$, $\bar{y} = 8$ also $\sigma_x = 8$, $\sigma_y = 2$

Now equation of regression of $x$ on $y$ is given by:

$$x - \bar{x} = \dfrac{r\,\sigma_x}{\sigma_y}(y - \bar{y})$$

$\Rightarrow x - 10 = \dfrac{(0.5)8}{2}(y - 8)$

$\Rightarrow x = 2y - 6$

$\therefore$ When rainfall is 9cm, production of wheat is estimated to be

$2(9) - 6 = 12$ kg. per unit area

**Example 22** Find the coefficient of correlation and the lines of regression for the data given below:

$n = 18, \sum x = 12, \sum y = 18, \sum x^2 = 60, \sum y^2 = 96$ and $\sum xy = 48$

**Solution**: $\bar{x} = \dfrac{\sum x}{n} = \dfrac{12}{18} = 0.67$, $\bar{y} = \dfrac{\sum y}{n} = \dfrac{18}{18} = 1$

$\sigma_x^2 = \dfrac{\sum x^2}{n} - \left(\dfrac{\sum x}{n}\right)^2 = \dfrac{60}{18} - \left(\dfrac{12}{18}\right)^2 = 2.89 \therefore \sigma_x = 1.7$

$\sigma_y^2 = \dfrac{\sum y^2}{n} - \left(\dfrac{\sum y}{n}\right)^2 = \dfrac{96}{18} - \left(\dfrac{18}{18}\right)^2 = 4.33 \therefore \sigma_y = 2.08$

$$r = \dfrac{\sum xy - \dfrac{\sum x \sum y}{n}}{\sqrt{(\sum x^2) - \dfrac{1}{n}(\sum x)^2}\sqrt{(\sum y^2) - \dfrac{1}{n}(\sum y)^2}}$$

$$= \frac{48 - \frac{(12)(18)}{18}}{\sqrt{(60) - \frac{1}{18}(12)^2}\sqrt{(96) - \frac{1}{18}(18)^2}} = \frac{36}{(7.2)(8.83)} = 0.57$$

$$b_{xy} = \frac{r\,\sigma_x}{\sigma_y} = \frac{(0.57)(1.7)}{2.08} = 0.47 \;,\; b_{yx} = \frac{r\,\sigma_y}{\sigma_x} = \frac{(0.57)(2.08)}{1.7} = 0.7$$

Equations of lines of regression are:

$$y - \bar{y} = b_{yx}(x - \bar{x}) \;,\; x - \bar{x} = b_{xy}(y - \bar{y})$$

$\Rightarrow y - 1 = 0.7(x - 0.67)$ and $x - 0.67 = 0.47(y - 1)$

$\Rightarrow y = 0.7x + 0.53$ and $x = 0.47y + 0.2$

**Example 23** Marks obtained by 11 students in statistics papers are given below:

| Paper I  | 60 | 65 | 68 | 70 | 75 | 85 | 80 | 45 | 55 | 56 | 58 |
| Paper II | 62 | 64 | 65 | 70 | 74 | 90 | 82 | 56 | 50 | 48 | 60 |

Calculate the coefficient of correlation and the equations of lines of regression.

**Solution**: Let marks obtained in paper I be denoted by $x$ and marks obtained in paper II be denoted by $y$.

Let $A_x = 65$, $A_y = 70$ $\therefore d_x = x - 65$, $d_y = y - 70$

Calculating $\sum d_x, \sum d_y, \sum d_x^2, \sum d_y^2$ and $\sum d_x d_y$

| $x$ | $d_x$ $(x-65)$ | $d_x^2$ | $y$ | $d_y$ $(y-70)$ | $d_y^2$ | $d_x d_y$ |
|---|---|---|---|---|---|---|
| 60 | -5 | 25 | 62 | -8 | 64 | 40 |
| 65 | 0 | 0 | 64 | -6 | 36 | 0 |
| 68 | 3 | 9 | 65 | -5 | 25 | -15 |
| 70 | 5 | 25 | 70 | 0 | 0 | 0 |
| 75 | 10 | 100 | 74 | 4 | 16 | 40 |
| 85 | 20 | 400 | 90 | 20 | 400 | 400 |
| 80 | 15 | 225 | 82 | 12 | 144 | 180 |
| 45 | -20 | 400 | 56 | -14 | 196 | 280 |
| 55 | -10 | 100 | 50 | -20 | 400 | 200 |
| 56 | -9 | 81 | 48 | -22 | 484 | 198 |
| 58 | -7 | 49 | 60 | -10 | 100 | 70 |
|    | $\sum d_x = 2$ | $\sum d_x^2 = 1414$ |  | $\sum d_y = -49$ | $\sum d_y^2 = 1865$ | $\sum d_x d_y = 1393$ |

Karl Pearson coefficient of correlation $(r)$ is given by:

$$r = \frac{\sum d_x d_y - \frac{\sum d_x \sum d_y}{n}}{\sqrt{(\sum d_x^2) - \frac{1}{n}(\sum d_x)^2} \sqrt{(\sum d_y^2) - \frac{1}{n}(\sum d_y)^2}}$$

$$\therefore r = \frac{1393 - \frac{(2)(-49)}{11}}{\sqrt{(1414) - \frac{1}{11}(2)^2} \sqrt{(1865) - \frac{1}{11}(-49)^2}} = \frac{1401.9091}{(37.5984)(40.5799)} = 0.9188$$

Now $\bar{x} = A_x + \frac{\sum d_x}{n} = 65 + \frac{2}{11} = 65.1818$

$\bar{y} = A_y + \frac{\sum d_y}{n} = 70 + \frac{-49}{11} = 65.5455$

Also $\sigma_x = \sqrt{\frac{\sum d_x^2}{n} - \left(\frac{\sum d_x}{n}\right)^2} = \sqrt{\frac{1414}{11} - \left(\frac{2}{11}\right)^2} = 11.3363$

and $\sigma_y = \sqrt{\frac{\sum d_y^2}{n} - \left(\frac{\sum d_y}{n}\right)^2} = \sqrt{\frac{1865}{11} - \left(\frac{-49}{11}\right)^2} = 12.2353$

$\therefore b_{xy} = \frac{r\,\sigma_x}{\sigma_y} = \frac{(0.9188)(11.3363)}{12.2353} = 0.8513$

$b_{yx} = \frac{r\,\sigma_y}{\sigma_x} = \frac{(0.9188)(12.2353)}{11.3363} = 0.9917$

Equations of lines of regression are:

$y - \bar{y} = b_{yx}(x - \bar{x})$ , $x - \bar{x} = b_{xy}(y - \bar{y})$

$\Rightarrow y - 65.55 = 0.99(x - 65.18)$ and $x - 65.18 = 0.85(y - 65.55)$

$\Rightarrow y = 0.99x + 1.02$  and  $x = 0.85y + 9.46$

**Example 24** The regression equations calculated from a given set of observations for two variables $x$ and $y$ are: $x = 9y + 5$ and $y = kx + 9$

Show that $0 < k < \frac{1}{9}$. Also if $k = \frac{1}{10}$, find $\bar{x}, \bar{y}$ and $r$

**Solution**: The two equations of regression are:

$x = 9y + 5$ ... ①

$y = kx + 9$ ... ②

$\Rightarrow b_{xy} = 9$ and $b_{yx} = k$

$\therefore r^2 = b_{xy} b_{yx} = 9k$

$\Rightarrow r = 3\sqrt{k}$ $\because b_{xy} = 9$ is positive, therefore $k$ and $r$ are also positive

Now $0 < r < 1$ or $0 < 3\sqrt{k} < 1$

$\Rightarrow 0 < 9k < 1$ or $0 < k < \frac{1}{9}$

Now if $k = \frac{1}{10}$, equation ② becomes $10y = x + 90$ ... ③

Solving ① and ③, the point of intersection of two regression lines is

$$\bar{x} = 860, \bar{y} = 95, \text{ also } r = 3\sqrt{k} = 3\sqrt{\frac{1}{10}} = 0.949$$

### Exercise 4

1. Fit a straight line $y = ax + b$ to the following data

| x | 0 | 1 | 3 | 6 | 8 |
|---|---|---|---|---|---|
| y | 1 | 3 | 2 | 5 | 4 |

2. Fit a straight line $y = a + bx$ to the following data

| x | 25 | 19 | 50 | 36 | 40 | 45 | 30 |
|---|----|----|----|----|----|----|----|
| y | 77 | 76 | 85 | 80 | 82 | 83 | 79 |

3. Fit a second degree parabola $y = ax^2 + bx + c$ to the following data

| x | 0 | 1 | 2 | 3 | 4 |
|---|---|---|---|---|---|
| y | 1 | 1.8 | 1.3 | 2.5 | 6.3 |

4. Fit a second degree parabola $y = a + bx + cx^2$ to the following data

| x | 2 | 4 | 6 | 8 | 10 |
|---|---|---|---|---|----|
| y | 3.07 | 12.85 | 31.47 | 57.38 | 91.29 |

5. Find the coefficient of correlation between $x$ and $y$ from the given data. Also find the two lines of regression.

| x | 1 | 2 | 3 | 4 | 5 | 6 | 7 | 8 | 9 | 10 |
|---|---|---|---|---|---|---|---|---|---|----|
| y | 10 | 12 | 16 | 28 | 25 | 36 | 41 | 49 | 40 | 50 |

6. Find the rank correlation for the following data:

| x | 56 | 42 | 72 | 36 | 63 | 47 | 55 | 49 | 38 | 42 | 68 | 60 |
|---|----|----|----|----|----|----|----|----|----|----|----|----|
| y | 147 | 125 | 160 | 118 | 149 | 128 | 150 | 145 | 115 | 140 | 152 | 155 |

7. Following table shows ages of husband and wife of 53 married couples.

| Wife→ <br> Husband↓ | 15-25 | 25-35 | 35-45 | 45-55 | 55-65 | 65-75 |
|---|---|---|---|---|---|---|
| 15-25 | 1 | 1 | - | - | - | - |
| 25-35 | 2 | 12 | 1 | - | - | - |
| 35-45 | - | 4 | 10 | 1 | - | - |
| 45-55 | - | - | 3 | 6 | 1 | - |
| 55-65 | - | - | - | 2 | 4 | 2 |
| 65-75 | - | - | - | - | 1 | 2 |

Calculate the coefficient of correlation between the age of the husband and that of wife.

8. The regression equations of two variables $x$ and $y$ are $x = 0.7y + 5.2$,

$y = 0.3x + 2.8$. Find the means of the two variables and the coefficient of correlation between them.

9. If the coefficient of correlation between two variables $x$ and $y$ is 0.5 and the acute angle between their lines of regression is $\tan^{-1}\frac{3}{8}$, show that $\sigma_x = \frac{\sigma_y}{2+\sqrt{3}}$

10. From a partially destroyed lab data, following results were retrieved:
    Lines of regression are:
    $$8x = 10y - 66 \text{ and } 18y = 40x - 214, \ \sigma_x = 3$$
    Find $\bar{x}, \bar{y}, \sigma_y$ and $r$ for the existing data.

## Answers

1. $y = 0.38x + 1.6$
2. $y = 70.052 + 0.292x$
3. $y = 0.55x^2 - 1.07x + 1.42$
4. $y = 0.34 - 0.78x + 0.99x^2$
5. $r = 0.96, x = 0.2y - 0.64, y = 4.69x + 4.9$
6. 0.932
7. 0.91
8. $\bar{x} = 9.06, \bar{y} = 5.52, r = 0.46$
10. $\bar{x} = 13, \bar{y} = 17, \sigma_y = 4, r = 0.6$

*Chapter 5: Sampling Distributions & Hypothesis Testing*

# Sampling Distributions And Hypothesis Testing

## 5.1 Introduction

Sampling is a statistical method of obtaining representative data (observations) from a group. We have been using sampling concepts in our day to day lives knowingly or unknowingly; for instance we take a handful of rice to check the rice quality of the full lot. This is an example of random sampling from a large population.

**Population (Universe):** The group of objects (individuals) under study is called population or universe. Universe may be finite or infinite.

**Sample:** A part containing objects (individuals), selected from the population is called a sample.

**Random Sampling:** The selection of objects (individuals) from the universe in such a way that each object (individual) of the universe has the same chance of

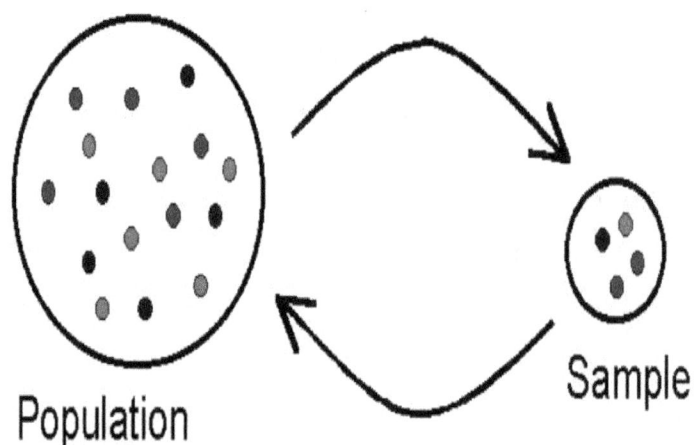

being selected is called random sampling. Lottery system is the most common example of random sampling.

**Simple Sampling:** Simple sampling is a special case of random sampling in which each event has same probability of success or failure.

**Note:** Every random sampling need not be simple. For example if balls are drawn without replacement from a bag of balls containing different balls; the probability of success changes in every trial. Thus the sampling though random is not simple.

**Hypothesis:** A hypothesis is an assumption based on insubstantial evidences that lends itself to further testing and experimentation. For example a farmer claims significant increase in crop production after using a particular fertilizer and after a season of experimenting, his hypothesis may be proved true or false. Any hypothesis may be accepted or rejected as per specific confidence levels and must be admissible to refutation.

**Null Hypothesis:** A hypothesis which is tested for possible rejection under the assumption of being true is known as null hypothesis. Usually the null hypothesis is stated as 'There is no relationship between two quantities'. It is denoted by $H_0$.

**Alternative Hypothesis:** It is the opposite statement of null hypothesis and denoted by $H_1$.

**Significance levels($\alpha$):** The probability levels below which we reject a hypothesis $H_0$ are called levels of significance. Most common significance levels employed in hypothesis testing are $\alpha = 5\%$, $\alpha = 1\%$, $\alpha = 0.27\%$ in which critical (rejection) regions occupy 5%, 1% and 0.27% areas of normal curve respectively.

**One Tailed and Two Tailed Tests:** While testing statistical significance levels; one-tailed test and a two-tailed test are used for accepting or rejecting a hypothesis. One-tailed tests are used for asymmetric distributions (reference value is unidirectional) which have a single tail; such as the chi-square distribution.

A two-tailed test is appropriate if the estimated value may lie on both sides of reference value. Two-tailed tests are only applicable when the probability curve has two tails; such as normal distribution.

One-tailed test; $\alpha = 1\%$    Two-tailed test; $\alpha = 1\%$    One-tailed test; $\alpha = 5\%$

Reference z values (two-tailed) for various levels are given in the table below:

| Significance level | Confidence level | z value |
|---|---|---|
| 0.27 % | 99.73% | 3.00 |
| 1.00% | 99.00% | 2.58 |
| 4.55% | 95.45% | 2.00 |
| 5.00% | 95.00% | 1.96 |
| 9.89% | 90.11% | 1.65 |

Testing at 1% level of confidence means that assumption has 99% chance of being true as rejection zone is just 1% of the total area under the normal curve. Similarly testing at 5% would provide 95% confidence level of acceptance.

**Type I and Type II Errors:** When we test a statistic at specified confidence level, there are chances of taking wrong decisions due to small sample size or sampling fluctuations etc.

Type I error is the incorrect rejection of a true null hypothesis, i.e. we reject $H_0$, when it is true.

Type II error is the incorrect acceptance of a false null hypothesis, i.e. we accept $H_0$, when it is false.

**Testing of Hypothesis:**

Testing of statistical hypothesis is a procedure designed for accepting or rejecting a hypothesis on the basis of some preset values.

**Step1:** Plant Null Hypothesis $H_0$ and alternate hypothesis $H_1$(optional) where $H_0$ is the hypothesis of no difference, i.e. $H_0$ presumes that there is no significant difference between observed value and expected value.

**Step2:** Find the most befitting test statistic for the analysis.

**Step3:** Take a random sample and compute the test statistic.

**Step 4:** $H_0$ is accepted if the value of test statistic lies in acceptance zone and rejected if it falls in critical (rejection) region at the desired significance level.

## 5.2 Sampling Distributions

A sampling distribution is a distribution of all of the possible values of a statistic; computed from randomly drawn samples of the same size from a population.

Some commonly used notations in sampling distributions are given below:

|  | Population | Sample |
|---|---|---|
| **Size** | $N$ | $n$ |
| **Mean** | $\mu$ | $\bar{x}$ |
| **Variance** | $\sigma^2$ | $s^2$ |
| **Standard Deviation** | $\sigma$ | $s$ |

Suppose we take various samples each of size $n$ from a population. If $p$ and $q$ be the probabilities of success and failure of each member of the sample, then the binomial distribution given by $(p + q)^n$ provides the sampling distribution of the number of successes in the sample with mean $np$ and variance $npq$.

- Mean (expected value) of number of successes $= np$
- Standard deviation $= \sqrt{npq}$.
- Probable occurrence range at 99.73% confidence level i.e. 0.27% significance level is given by: $np \pm 3\sqrt{npq}$
- Probable occurrence range at 99% confidence level i.e. 1% significance level is given by: $np \pm 2.58\sqrt{npq}$
- Probable occurrence range at 95% confidence level i.e. 5% significance level is given by: $np \pm 1.96\sqrt{npq}$

In case of proportion of successes, mean and standard deviation of proportion of successes are obtained by dividing each statistic by $n$.

- Mean (expected value) of proportion of successes $= \frac{np}{n} = p$
- Standard deviation $= \frac{\sqrt{npq}}{n} = \sqrt{\frac{pq}{n}}$
- Probable occurrence range of the proportion at 99.73% confidence level i.e. 0.27% significance level is given by: $p \pm 3\sqrt{\frac{pq}{n}}$
- Probable occurrence range of the proportion at 99% confidence level i.e. 1% significance level is given by: $p \pm 2.58\sqrt{\frac{pq}{n}}$
- Probable occurrence range of the proportion at 95% confidence level i.e. 5% significance level is given by: $p \pm 1.96\sqrt{\frac{pq}{n}}$

**Standard Error:** The standard deviation of the sampling distribution of a statistic is known as Standard Error (S.E.).

**Precision**: Reciprocal of standard error is known as precision.

**Probable Error**: It is taken as 0.67449 times the standard error and is used sometimes to explain the concept of sampling errors to layman or unprofessional people.

## 5.3 Sampling of Attributes for large samples ($n > 30$)

Characteristics like language, religion, habits (traits) etc. cannot be measured in numbers as they are attributes. Sampling of attributes means testing how many in a population possess a particular attribute (trait) or whether the two populations share an attribute (trait) in common and to how much confidence level.

When sample size ($n$) is very large i.e. greater than 30 and neither $p$ nor $q$ are very small, the binomial distribution tends to normal distribution and therefore we choose the variate $z$ as test statistic.

Following procedure is adopted for testing the significance of large samples in terms of attributes.

**Step1**: Postulate the null hypothesis ($H_0$); if required.

**Step2**: If $x$ is the observed number of successes in a sample and $z$ is the standard normal variate, then $z = \frac{x-\mu}{\sigma}$, i.e. $z = \frac{x-np}{\sqrt{npq}}$

**Step3**: Accept or reject $H_0$ as per given values of parameter $z$ at different significance levels:

| Significance Level | $H_0$ accepted | $H_0$ rejected |
|---|---|---|
| 0.27% | $|z| < 3$ | $|z| > 3$ |
| 1% | $|z| < 2.58$ | $|z| > 2.58$ |
| 5% | $|z| < 1.96$ | $|z| > 1.96$ |

In case of absence of any specified significance level, we may consider 0.27% level, i.e. take acceptance range as $-3 < z < 3$.

**Example 1** A coin is tossed 400 times and turns up head 216 times. Discuss whether the coin may be unbiased one.

**Solution**: Let $H_0$: coin is unbiased

Here $n = 400$, if $p$ denotes probability of success; i.e. getting a head, then $p = \frac{1}{2}$, $q = \frac{1}{2}$, ∴ Expected number of heads is $np = 400 \times \frac{1}{2} = 200$

Again if $x$ denotes observed number of heads; then the test statistic

$$z = \frac{x-np}{\sqrt{npq}}; \text{i.e. } z = \frac{216-200}{\sqrt{400 \times \frac{1}{2} \times \frac{1}{2}}} = 1.6 < 1.96 < 2.58$$

∴ $H_0$ is acceptable at both 5% and 1% levels of significance.

Hence our assumption that coin is unbiased is true

**Example2** A die is rolled 900 times and turns up five or six 338 times. Discuss whether the die may be unbiased one.

**Solution**: Let $H_0$: die is unbiased

Here $n = 900$, if $p$ denotes probability of success; i.e. getting 5 or 6, then $p = \frac{1}{3}$, $q = \frac{2}{3}$, $\therefore$ Expected number of successes is $np = 900 \times \frac{1}{3} = 300$

Again if $x$ denotes observed number of successes; then the test statistic

$$z = \frac{x - np}{\sqrt{npq}} \text{ ; i.e. } z = \frac{338 - 300}{\sqrt{900 \times \frac{1}{3} \times \frac{2}{3}}} = 2.69$$

$\therefore$ $2.69 < 3$   but $2.69 > 1.96$ and also $2.69 > 2.58$

Hence $H_0$ is acceptable at 0.27% level of significance but fails at both 1% and 5% significance levels.

**Example3** A random sample of 500 oranges was taken from a large consignment and 65 were found to be bad. Show that the standard error of the proportion of bad ones in a sample of this size is 0.015 and deduce that the percentage of bad oranges in the consignment lies between 8.5 and 17.5.

**Solution:** Let $p$ denote the proportion of bad oranges in the given sample

Then $p = \frac{65}{500} = 0.13$, $q = 1 - p = 0.87$

Standard Deviation (S.D.) $= \sqrt{\frac{pq}{n}} = \sqrt{\frac{0.13 \times 0.87}{500}} = 0.015$

$\therefore$ Standard Error (S.E.) = S.D. = 0.015

Probable limit of bad oranges in the consignment is given by:

$$p \pm 3\sqrt{\frac{pq}{n}} = 0.13 \pm 3(0.015)$$

$$= 0.085 \text{ to } 0.175$$

$\therefore$ Probable percentage of bad oranges in the consignment is 8.5% to 17.5%

**Example4** A random sample of 100 bolts was taken from the lot manufactured by a machine and 10 were found to be defective. Find the 95% confidence limits for the proportion of defective bolts produced by the machine.

**Solution:** Let $p$ denote the proportion of defective bolts in the given sample

Then $p = \frac{10}{100} = 0.1$, $q = 1 - p = 0.9$

Standard Deviation (S.D.) $= \sqrt{\frac{pq}{n}} = \sqrt{\frac{0.1 \times 0.9}{100}} = 0.03$

Probable limit of defective bolts in the lot at 95% confidence level is given by:

$$p \pm 1.96\sqrt{\frac{pq}{n}} = 0.1 \pm 1.96(0.03) = 0.1 \pm 0.0588 = 0.0412 \text{ to } 0.1588$$

$\therefore$ Probable percentage of proportion of defective bolts in the lot at 95% confidence level is 4.12% to 15.88%

**Example5** A sample of 900 days is taken from metrological records of a district and 100 of them are found to be foggy. What is the probable percentage of foggy days in the district?

**Solution:** Let $p$ denote the probability of a foggy day in the district, then

$$p = \frac{100}{900} = \frac{1}{9}, q = 1 - p = \frac{8}{9}$$

$$\text{Standard Deviation (S.D.)} = \sqrt{npq} = \sqrt{900 \times \frac{1}{9} \times \frac{8}{9}} = 9.43$$

Probable limit of foggy days at 99.73% confidence level is given by:

$$np \pm 3\sqrt{npq} = 100 \pm 3(9.43) = 100 \pm 28.29 \text{; i.e. } 71.71 \text{ to } 128.29$$

Probable percentage is $\frac{71.71}{900} \times 100$ to $\frac{128.29}{900} \times 100$ i.e. 7.97% to 14.25%

## 5.3.1 Comparing Proportions of Large Samples from Two Different Populations in Terms of Attributes

For an attribute (trait), let proportions $p_1$ and $p_2$ be given from two large samples of sizes $n_1$ and $n_2$ respectively from two different populations. We may want to test whether two populations are similar regarding the specified attribute (trait).

Working methodology:
1. Set up the hypothesis $H_0$: The two populations are similar regarding the specified attribute (trait)
2. Find common proportion of two populations for the specified attribute as: $p = \frac{n_1 p_1 + n_2 p_2}{n_1 + n_2}$, $q = 1 - p$
3. Compute combined standard error of two populations as: $e = \sqrt{pq \left(\frac{1}{n_1} + \frac{1}{n_2}\right)}$
4. Compute the statistic parameter $z = \frac{p_1 - p_2}{e}$
5. Accept or reject $H_0$ as per given values of parameter $z$ at different significance levels:

| Significance Level | $H_0$ accepted | $H_0$ rejected |
|---|---|---|
| 0.27% | $|z| < 3$ | $|z| > 3$ |
| 1% | $|z| < 2.58$ | $|z| > 2.58$ |
| 5% | $|z| < 1.96$ | $|z| > 1.96$ |

**Example 6** In a sample of 600 men from a certain city, 450 are found to be smokers. In another sample of 700 men from another city, 450 are found smokers. Do the data indicate that the cities are significantly different with respect to the habit of smoking among men?

**Solution**: Let $H_0$: The two populations are similar regarding smoking habits among men.

Here $n_1 = 600$, $p_1 = \frac{450}{600} = 0.75$

$n_2 = 900$, $p_2 = \frac{450}{700} = 0.64$

$p = \frac{n_1 p_1 + n_2 p_2}{n_1 + n_2} = \frac{600 \times 0.75 + 700 \times 0.64}{600 + 700} = \frac{449}{650} = 0.69$

$q = 1 - p = 1 - 0.69 = 0.31$

$\therefore e = \sqrt{pq \left(\frac{1}{n_1} + \frac{1}{n_2}\right)} = \sqrt{0.69 \times 0.31 \left(\frac{1}{600} + \frac{1}{700}\right)} = 0.026$

$z = \frac{p_1 - p_2}{e} = \frac{0.75 - 0.64}{0.026} = 4.23$

$|z| > 3$, $\therefore H_0$ is rejected, i.e. the difference between two populations is highly significant in terms of smoking habits among men both at 1% and 5% levels of significance.

**Example 7** In a large city $A$, 20% of a random sample of 900 school boys had defective eye-sight. In another large city $B$, 15.5% of a random sample of 1600 school boys had defective eye-sight. Is the difference between two proportions significant?

**Solution**: Let $H_0$: The two populations are similar regarding school boys having defective eye sight.

Here $n_1 = 900$, $p_1 = \frac{20}{100} = \frac{1}{5} = 0.2$

$n_2 = 1600$, $p_2 = \frac{15.5}{100} = \frac{31}{200} = 0.155$

$p = \frac{n_1 p_1 + n_2 p_2}{n_1 + n_2} = \frac{900 \times 0.2 + 1600 \times 0.155}{900 + 1600} = 0.1712$

$q = 1 - p = 1 - 0.1712 = 0.8288$

$\therefore e = \sqrt{pq\left(\frac{1}{n_1} + \frac{1}{n_2}\right)} = \sqrt{0.1712 \times 0.8288 \left(\frac{1}{900} + \frac{1}{1600}\right)} = 0.01568$

$z = \frac{p_1 - p_2}{e} = \frac{0.2 - 0.155}{0.01568} = 2.8699$

$|z| < 3$, $\therefore H_0$ is accepted at 0.27% level of significance, i.e. The difference between two populations is not significant in terms of defective eye-sights.

**Remark**: $H_0$ should be rejected at 1% or 5% significance levels.

**Example 8** Following data gives proportion of dark coloured people in two cities.

| City | Sample size | Percentage of dark coloured people |
|------|-------------|-----------------------------------|
| A    | 250         | 42                                |
| B    | 450         | 48                                |

Can the difference between two percentages taken as sampling fluctuations?

**Solution**: Let $H_0$: The two populations are similar regarding smoking habits among men.

Here $n_1 = 250$, $p_1 = 42\%$

$n_2 = 450$, $p_2 = 48\%$

$p = \frac{n_1 p_1 + n_2 p_2}{n_1 + n_2} = \frac{250 \times 42 + 450 \times 48}{250 + 450} = 45.86\%$

$q = 1 - p = 54.14\%$

$\therefore e = \sqrt{pq\left(\frac{1}{n_1} + \frac{1}{n_2}\right)} = \sqrt{45.86 \times 54.14 \left(\frac{1}{250} + \frac{1}{450}\right)} = 3.93\%$

$z = \frac{p_1 - p_2}{e} = \frac{42 - 48}{3.93} = -1.53$

$|z| < 1.96$, $\therefore H_0$ is accepted, i.e. The difference between two populations is not significant in terms of dark coloured people both at 1% and 5% levels of significance.

## 5.4 Sampling Distributions of Sample means

The sampling distribution of the mean refers to the pattern of sample means, observed by different samples drawn from the population at large.

**Result I** : If all possible samples of size $n$ are drawn without replacement from a finite population of size $N$ and if $\mu$ and $\sigma$ denote population mean and standard deviation respectively;

$\mu_{\bar{x}}$ and $\sigma_{\bar{x}}$ denote mean and standard deviation respectively of sampling distribution, then $\mu_{\bar{x}} = \mu$ and $\sigma_{\bar{x}} = \frac{\sigma}{\sqrt{n}} \sqrt{\frac{N-n}{N-1}}$

**Result II**: If all possible samples of size $n$ are drawn with replacement from a finite population of size $N$ and if $\mu$ and $\sigma$ denote population mean and standard deviation respectively;

$\mu_{\bar{x}}$ and $\sigma_{\bar{x}}$ denote mean and standard deviation respectively of sampling distribution, then $\mu_{\bar{x}} = \mu$ and $\sigma_{\bar{x}} = \frac{\sigma}{\sqrt{n}}$

**Example 9** Suppose a population consists of five numbers: 1,3,5,7,9. Find sampling distribution of sample means. Also calculate mean and standard deviation of the sample means, if random samples of two numbers are drawn

i. without replacement ii. with replacement

Verify Results *I* and *II* by comparing these statistics with population mean and population standard deviation.

**Solution**: Population mean $(\mu) = \frac{1}{N}\sum_1^N x_i$

$\mu = \frac{1}{5}[1 + 3 + 5 + 7 + 9] = 5$

Population variance $(\sigma^2) = \frac{1}{N}\sum_1^N (x_i - \mu)^2$

$\sigma^2 = \frac{1}{5}[(1-5)^2 + (3-5)^2 + 0 + (7-5)^2 + (9-5)^2] = \frac{40}{5} = 8$, $\therefore \sigma = \sqrt{8}$

i. Total number of possible samples each of size two without replacement is $^5C_2 = 10$

| Sample | Sample mean $(\bar{x})$ |
|---|---|
| 1,3 | 2 |
| 1,5 | 3 |
| 1,7 | 4 |
| 1,9 | 5 |
| 3,5 | 4 |
| 3,7 | 5 |
| 3,9 | 6 |
| 5,7 | 6 |
| 5,9 | 7 |
| 7,9 | 8 |

Sampling distribution of Sample mean $(\bar{x})$

| $\bar{x}$ | 2 | 3 | 4 | 5 | 6 | 7 | 8 |
|---|---|---|---|---|---|---|---|
| Frequency | 1 | 1 | 2 | 2 | 2 | 1 | 1 |

Mean of sample means $(\mu_{\bar{x}}) = \frac{1}{n}\sum_1^n \bar{x}$

$= \frac{1}{10}[2 + 3 + 2(4) + 2(5) + 2(6) + 7 + 8] = 5$

Variance of sample means

$\sigma_{\bar{x}}^2 = \frac{1}{10}[(2-5)^2 + (3-5)^2 + 2(4-5)^2 + 0 + 2(6-5)^2 + (7-5)^2 + (8-5)^2] = \frac{30}{10} = 3$

Standard deviation of sample means $(\sigma_{\bar{x}}) = \sqrt{3}$

$\therefore \mu_{\bar{x}} = 5 = \mu$, also $\frac{\sigma}{\sqrt{n}} \sqrt{\frac{N-n}{N-1}} = \frac{\sqrt{8}}{\sqrt{2}} \sqrt{\frac{5-2}{5-1}} = \sqrt{3} = \sigma_{\bar{x}}$

Hence Result *I* is verified.

ii. Total number of possible samples each of size two with replacement is $5 \times 5 = 25$

Samples of 2 numbers with replacement are shown below; individual sample mean is given under each sample.

| (1,1) | (1,3) | (1,5) | (1,7) | (1,9) |
|---|---|---|---|---|
| 1 | 2 | 3 | 4 | 5 |
| (3,1) | (3,3) | (3,5) | (3,7) | (3,9) |
| 2 | 3 | 4 | 5 | 6 |
| (5,1) | (5,3) | (5,5) | (5,7) | (5,9) |
| 3 | 4 | 5 | 6 | 7 |
| (7,1) | (7,3) | (7,5) | (7,7) | (7,9) |
| 4 | 5 | 6 | 7 | 8 |
| (9,1) | (9,3) | (9,5) | (9,7) | (9,9) |
| 5 | 6 | 7 | 8 | 9 |

Mean of sample means is given by:

$\mu_{\bar{x}} = \frac{1}{25}[1 + 2(2) + 3(3) + 4(4) + 5(5) + 4(6) + 3(7) + 2(8) + 9] = 5$

Variance of sample means is given by:

$\sigma_{\bar{x}}^2 = \frac{1}{25}[(1-5)^2 + 2(2-5)^2 + 3(3-5)^2 + 4(4-5)^2 + 5(5-5)^2 +$
$\qquad 4(6-5)^2 + 3(7-5)^2 + 2(8-5)^2 + (9-5)^2]$
$= \frac{1}{25}[16 + 18 + 12 + 4 + 0 + 4 + 12 + 18 + 16] = \frac{100}{25} = 4$

$\therefore$ Standard deviation of sample means ($\sigma_{\bar{x}}$) = 2

$\mu_{\bar{x}} = 5 = \mu$ and $\sigma_{\bar{x}} = 2 = \frac{8}{\sqrt{2}} = \frac{\sigma}{\sqrt{n}}$, hence Result *II* is verified.

**Result III**: If a population is normally distributed with mean and variance $\mu$ and $\sigma^2$ respectively; also $\bar{x}$ denotes sample mean for a random sample $(x_1, x_2, \cdots, x_n)$; then sampling distribution of $\bar{x}$ is also normally distributed with mean $\mu$ and variance $\frac{\sigma^2}{n}$, i.e. If $x \sim N(\mu, \sigma^2)$ then $\bar{x} \sim N\left(\mu, \frac{\sigma^2}{n}\right)$

**Proof**: Since the population is normally distributed, for any object $x_i$ of the population;
$E(x_i) = \mu$ and $Var(x_i) = \sigma^2$

$\therefore E(\bar{x}) = E\left(\frac{x_1 + x_2 + \cdots + x_n}{n}\right)$

$= E\left(\frac{x_1}{n}\right) + E\left(\frac{x_2}{n}\right) + \cdots + E\left(\frac{x_n}{n}\right)$

$= \frac{1}{n}\mu + \frac{1}{n}\mu + \cdots + \frac{1}{n}\mu = \frac{n\mu}{n} = \mu$

Also $Var(\bar{x}) = Var\left(\frac{x_1 + x_2 + \cdots + x_n}{n}\right)$

$= Var\left(\frac{x_1}{n}\right) + Var\left(\frac{x_2}{n}\right) + \cdots + Var\left(\frac{x_n}{n}\right)$

$$= \frac{1}{n^2}\sigma^2 + \frac{1}{n^2}\sigma^2 + \cdots + \frac{1}{n^2}\sigma^2 = \frac{n\sigma^2}{n^2} = \frac{\sigma^2}{n}$$

$$\because Var(ax_i) = a^2 Var(x_i)$$

∴ Standard deviation of sampling distribution ($\sigma_{\bar{x}}$) $= \frac{\sigma}{\sqrt{n}}$

**Standard Error**: The standard deviation of the sampling distribution is called the standard error

∴ Standard error of sampling distribution is $\sigma_{\bar{x}} = \frac{\sigma}{\sqrt{n}}$

**Remark**: We have assumed here that $\sigma$ is known. However if $\sigma$ is not known, we take $\sigma$ to be equal to the standard deviation of the sample.

**Properties of the sampling distribution of sample means:**

➢ A sample drawn from a normally distributed population follows normal distribution,

∴ Z-value for the distribution of $\bar{x}$ is given by $\frac{\bar{x}-\mu}{\sigma_{\bar{x}}} = \frac{\bar{x}-\mu}{\frac{\sigma}{\sqrt{n}}}$

➢ The mean $\mu_{\bar{x}}$ of the sample means will be the same as population mean from which the samples were drawn, i.e. $\mu_{\bar{x}} = \mu$

➢ The variance $\sigma_{\bar{x}}^2$ of the sampling distribution of $\bar{x}$ will be equal to the variance of the population divided by the sample size i.e. $\sigma_{\bar{x}}^2 = \frac{\sigma^2}{n}$

**Result IV : Central Limit Theorem:** As the sample size gets large enough (30 or higher); the sampling distribution becomes approximately normal regardless of shape of population.

**Remark:** For large samples ($n \geq 30$), probability distribution is taken as normal for computational purposes.

**Example 10** A population has mean 0.1 and standard deviation 2.1. Find the probability that the mean of a random sample of size 900 will be negative.

**Solution**: Given that population mean ($\mu$) = 0.1 and standard deviation ($\sigma$) = 2.1

Since the sample size is large enough, sampling distribution is approximately normal with mean 0.1 and standard deviation $= \frac{\sigma}{\sqrt{n}} = \frac{2.1}{\sqrt{900}} = 0.07$

i.e. $\bar{x} \sim N\left(0.1, \frac{(2.1)^2}{900}\right)$ and $z = \frac{\bar{x}-\mu}{\frac{\sigma}{\sqrt{n}}} = \frac{\bar{x}-0.1}{\frac{2.1}{30}}$ $\Rightarrow \bar{x} = 0.1 + 0.07z$

$\therefore P(\bar{x} < 0) = P(0.1 + 0.07z < 0)$

$= P\left(z < \frac{-0.1}{0.07}\right)$

$= P(z < -1.43)$

$= P(z > 1.43)$

$= 0.5 - 0.4236 = 0.0764$

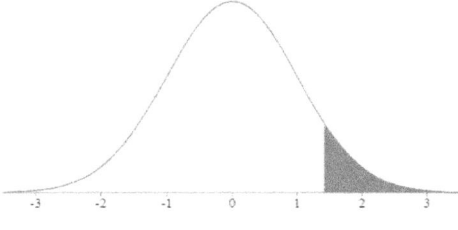

**Example 11** Suppose a population has mean 10 and variance 4. What is the probability that the sample of size 36 has mean lying between 9.8 and 10.2?

**Solution**: Given that population mean ($\mu$) = 10 and standard deviation ($\sigma$) = 2. Since the sample size is large enough, sampling distribution is approximately normal with mean 10 and standard deviation $= \frac{\sigma}{\sqrt{n}} = \frac{2}{\sqrt{36}} = 0.33$

i.e. $\bar{x} \sim N\left(10, \frac{4}{36}\right)$ and $z = \frac{\bar{x}-\mu}{\frac{\sigma}{\sqrt{n}}} = \frac{\bar{x}-10}{\frac{2}{6}}$ $\Rightarrow \bar{x} = 10 + 0.33z$

$\therefore P(9.8 < \bar{x} < 10.2)$
$= P(9.8 < 10 + 0.33z < 10.2)$
$= P\left(\frac{9.8-10}{0.33} < z < \frac{10.2-10}{0.33}\right)$
$= P(-0.61 < z < 0.61)$
$= 0.2291 + 0.2291 = 0.4582$

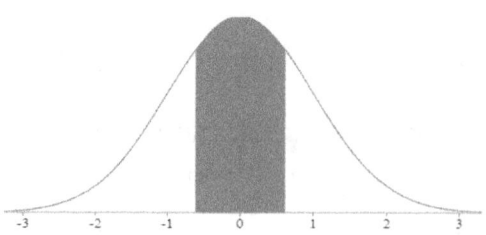

**Example 12** A firm produces electric bulbs that have normally distributed mean burning life 800 hours with a standard deviation of 40 hours. Find the probability that a random sample of 16 bulbs will have average burning life of less than 775 hours.

**Solution**: Given distribution is normal with population mean $(\mu) = 800$ and standard deviation $(\sigma) = 40$, sample size $(n) = 16$

$z = \frac{\bar{x}-\mu}{\frac{\sigma}{\sqrt{n}}} = \frac{\bar{x}-800}{10}$

$\Rightarrow \bar{x} = 800 + 10z$

$\therefore P(\bar{x} < 775) = P(800 + 10z < 775)$
$= P\left(z < \frac{775-800}{10}\right)$
$= P(z < -2.5)$
$= P(z > 2.5)$  By symmetry of standard normal curve
$= 0.5 - P(0 < z < 2.5) = 0.5 - 0.4938 = 0.0062$

**Result V : Test of significance of sample mean for a large sample**

Here we test the hypotheses about the sample mean of a population in the case when sample size is at least 30 (for large samples, distribution is approximately normal as per central limit theorem). Let $\mu$ and $\sigma$ be mean and variance of a population. If $\bar{x}$ be the mean of a simple large sample of size $n$ from an unknown population, we may want to test whether the sample belongs to given population.

Working rule:
1. Set up the hypothesis $H_0$: the sample belongs to given population
2. Compute the statistic parameter $z = \frac{\bar{x}-\mu}{\frac{\sigma}{\sqrt{n}}}$
3. Accept or reject $H_0$ as per given values of parameter $z$ at different significance levels:

| Significance Level | $H_0$ accepted | $H_0$ rejected |
|---|---|---|
| 0.27% | $|z| < 3$ | $|z| > 3$ |
| 1% | $|z| < 2.58$ | $|z| > 2.58$ |
| 5% | $|z| < 1.96$ | $|z| > 1.96$ |

**Remark:** The statistic parameter $z = \frac{\bar{x}-\mu}{\frac{\sigma}{\sqrt{n}}}$ can also be used to check whether the sample taken from the given population is random or not.

**Example 13** A sample of size 900 is having mean 3.6 mm; could it be reasonably regarded as a random sample from large population whose mean is 3.35 mm and standard deviation 2.6 mm at 1% significance level.

**Solution**: Let $H_0$: sample belongs to the given population

Here population mean $\mu = 3.35$ mm, population standard deviation $\sigma = 2.6$ mm,

also sample size $n = 900$ and sample mean $\bar{x} = 3.6$ mm

Now $z = \dfrac{\bar{x} - \mu}{\frac{\sigma}{\sqrt{n}}} = \dfrac{3.6 - 3.35}{\frac{2.6}{\sqrt{900}}} = \dfrac{3.6 - 3.35}{2.6}(30) = 2.88 > 2.58$

As $|z| > 2.58$, $H_0$ is rejected, i.e. difference between sample mean ($\bar{x}$) and population mean ($\mu$) is significant at 1% level and the sample may not be considered from the same population.

**Example 14** A random sample of 400 canes of vegetable oil with labeled net weight as 5kg has a mean net weight of 4.98 kg with a standard deviation of 0.22 kg. Can we accept the hypothesis of net weight 5 kg per cane on the basis of given sample, at 5 % level of significance?

**Solution**: Let $H_0$: Mean net weight of each tin is 5 kg.

Here population mean $\mu = 5$ kg, sample size $n = 400$ and sample mean $\bar{x} = 4.98$ kg,

sample standard deviation $s = 0.22$ kg

Now $z = \dfrac{\bar{x} - \mu}{\frac{\sigma}{\sqrt{n}}} = \dfrac{4.98 - 5}{\frac{0.22}{\sqrt{400}}} = \dfrac{4.98 - 5}{0.22}(20) = -1.81$

As $|z| < 1.96$, $H_0$ is accepted at 5% significance level, i.e. Mean net weight of each tin is 5 kg.

Note: Here population standard deviation $\sigma$ is not known, $\therefore$ sample standard deviation is taken as population standard deviation.

**Example 15** A pharmaceutical company fills its best-selling 8 ounce jars of liquid medicine by an automatic dispensing machine. The machine is set to dispense a mean of 8.1 ounces per jar with a standard deviation of 0.22 ounce. Uncontrollable factors in the process can shift the mean away from 8.1 and cause either under-fill or overfill, both of which are undesirable. In such a case the dispensing machine is stopped and recalibrated. A quality control engineer routinely selects 30 jars from the assembly line to check the amounts filled. On one occasion, the sample mean is 8.2 ounces with the standard deviation of 0.25 ounce. Determine if there is sufficient evidence in the sample to indicate, that the machine should be recalibrated, at the 5% level of significance.

**Solution**: Let $H_0$: machine need not be recalibrated

Here population mean $\mu = 8.1$ ounce, population standard deviation

$\sigma = 0.22$ ounce.

Also sample size $n = 30$ and sample mean $\bar{x} = 8.2$ ounce

$\therefore z = \dfrac{\bar{x} - \mu}{\frac{\sigma}{\sqrt{n}}} = \dfrac{8.2 - 8.1}{\frac{0.22}{\sqrt{30}}} = \dfrac{8.2 - 8.1}{0.22}(\sqrt{30}) = 0.45 \times 5.48 = 2.47$

As $|z| > 1.96$, $H_0$ is rejected, i.e. difference between sample mean ($\bar{x}$) and population mean ($\mu$) is significant at 5% level and the machine needs to be recalibrated.

**Result VI : Test of significance of difference between means of two large samples**

  **(a) If samples are from two different populations**

Let $\bar{x}_1$ and $\bar{x}_2$ be sample means of two populations of sizes $n_1$ and $n_2$ from two different normally distributed populations having means $\mu_1$; $\mu_2$ and variances $\sigma_1^2$ ; $\sigma_2^2$ respectively;

$$\text{i.e. } \bar{x}_1 \sim N\left(\mu_1, \frac{\sigma_1^2}{n_1}\right) \text{ and } \bar{x}_2 \sim N\left(\mu_2, \frac{\sigma_2^2}{n_2}\right)$$

Then $E(\bar{x}_1 - \bar{x}_2) = E(\bar{x}_1) - E(\bar{x}_2) = \mu_1 - \mu_2$

Also $Var(\bar{x}_1 - \bar{x}_2) = Var(\bar{x}_1) + Var(\bar{x}_2) = \frac{\sigma_1^2}{n_1} + \frac{\sigma_2^2}{n_2}$

∴ Standard Error $(e) = \sqrt{\frac{\sigma_1^2}{n_1} + \frac{\sigma_2^2}{n_2}}$

z-value for the distribution of $\bar{x}_1 - \bar{x}_2$ is given by $z = \frac{(\bar{x}_1 - \mu_1) - (\bar{x}_2 - \mu_2)}{e}$

  **(b) If samples are from same population**

If $\bar{x}_1$ and $\bar{x}_2$ be two sample means of sizes $n_1$ and $n_2$ taken from a normally distributed population having mean $\mu$ and variance $\sigma^2$

Then $E(\bar{x}_1 - \bar{x}_2) = E(\bar{x}_1) - E(\bar{x}_2) = \mu - \mu = 0$

Also $Var(\bar{x}_1 - \bar{x}_2) = Var(\bar{x}_1) + Var(\bar{x}_2) = \frac{\sigma^2}{n_1} + \frac{\sigma^2}{n_2}$

∴ Standard Error $(e) = \sqrt{\frac{\sigma^2}{n_1} + \frac{\sigma^2}{n_2}}$

z-value for the distribution of $\bar{x}_1 - \bar{x}_2$ is given by $z = \frac{\bar{x}_1 - \bar{x}_2}{e}$

**Example 16** A random sample of 150 villages was taken from a district A having standard deviation 32 and average population per village was found to be 440. Another random sample of 250 villages from district B with a standard deviation of 56 gave an average population of 480 per village. Is the difference between the averages of two populations significant? Give reasons.

**Solution**: Let $H_0$: The differences between averages of two populations is not significant,

  i.e. $\mu_1 = \mu_2$

Here $n_1 = 150$, $\bar{x}_1 = 440$, $\sigma_1 = 32$

$n_2 = 250$, $\bar{x}_2 = 480$, $\sigma_2 = 56$

$e = \sqrt{\frac{\sigma_1^2}{n_1} + \frac{\sigma_2^2}{n_2}} = \sqrt{\frac{32^2}{150} + \frac{56^2}{250}} = 4.4$

∴ $z = \frac{(\bar{x}_1 - \mu_1) - (\bar{x}_2 - \mu_2)}{e} = \frac{(\bar{x}_1 - \bar{x}_2) - (\mu_1 - \mu_2)}{e}$

$= \frac{440 - 480}{4.4} = -9.09$, under the assumption $\mu_1 = \mu_2$

$|z| \gg 3$, thus differences between two averages is highly significant.

∴ $H_0$ is rejected and the averages of two populations cannot be taken as same.

**Example 17** Random samples of 500 and 400 are having means 11.5 and 10.9 respectively. Can the two samples be regarded as drawn from the population of standard deviation 5?

**Solution**: Let $H_0$ : The two samples are drawn from the same population with standard deviation $(\sigma) = 5$

Here $n_1 = 500$, $\bar{x}_1 = 11.5$
$n_2 = 400$, $\bar{x}_2 = 10.9$

$$e = \sqrt{\frac{\sigma^2}{n_1} + \frac{\sigma^2}{n_2}} = \sqrt{\frac{25}{500} + \frac{25}{400}} = 0.335$$

$$\therefore z = \frac{\bar{x}_1 - \bar{x}_2}{e} = \frac{11.5 - 10.9}{0.335} = 1.79$$

$|z| < 1.96$, $\therefore H_0$ is accepted at 5% level of significance, i.e. the two samples can be considered to be drawn from same population.

**Example 18** 50 new entrants in a class are found to have an average height of 135cm and 30 old one have an average height of 140cm with a class standard deviation of 8cm. Does this indicate that mean height of old students is greater than that of new entrants?

**Solution**: Let $H_0$: Average height of new entrants is same as old students

Here standard deviation of the class is $\sigma = 8$

$n_1 = 50$, $\bar{x}_1 = 135$
$n_2 = 30$, $\bar{x}_2 = 140$,

$$e = \sqrt{\frac{\sigma^2}{n_1} + \frac{\sigma^2}{n_2}} = \sqrt{\frac{64}{50} + \frac{64}{30}} = 1.85$$

$$\therefore z = \frac{\bar{x}_1 - \bar{x}_2}{e} = \frac{135 - 140}{1.85} = -2.7$$

$|z| < 3$, $\therefore H_0$ is accepted at 0.27% significance level, i.e. average height of new entrants is same as old students.

Remark: $H_0$ will have to be rejected at 1% and 5% levels of significance.

## Exercise 5

1. A coin is tossed 400 times and head turns up 225 times. Discuss whether the coin is biased or unbiased at 5% level of significance.
2. A random sample of 600 oranges was taken from a large consignment and 60 were found to be rotten. Show that the standard error of the proportion of bad ones in a sample of this size is 0.1 and deduce that the percentage of bad oranges in the consignment almost lies between 6.3 and 13.7
3. In a city $A$ 20% of a random sample of 900 school children wore spectacles and in another city $B$ 18.5% of a random sample of 1600 school children used to wear spectacles. Is the difference between the proportions significant?
4. In a sample of 500 people from a state 280 take tea and rest take coffee. Can we assume that tea and coffee are equally popular in the state?

5. A sample of 900 members is found to have a mean of 3.4cm. Can it be reasonably regarded as truly random sample from a large population with mean 3.25cm and S.D. 1.61cm.
6. A sample of 100 electric bulbs produced by a manufacturer $M_1$ showed a mean life time 1190 hours with a standard deviation of 90 hours. Another sample of 75 bulbs produced by manufacturer $M_2$ showed a mean life time 1230 hours with a standard deviation of 120 hours. Is there a difference between the mean life times of two brands at 5% level of significance?
7. The means of two large samples of 1000 and 2000 members are 168.75 cm and 170cm respectively. Can these be regarded as drawn from the same population of standard deviation 6.25 cm.
8. A stenographer states that he can take dictation at the rate of 120 words per minute. Can we accept his claim on the basis of 100 trials in which he showed a mean of 116 words with standard deviation of 15 words ?
9. A sample of height of 6400 soldiers has a mean of 67.85 inches and a standard deviation of 2.56 inches, while a random sample of heights of 1600 sailors has a mean of 68.55 inches and a standard deviation of 2.52 inches. Does this indicate that the sailors are on average taller than the soldiers?
10. A random sample of 400 students has an average weight of 55 kg. Can we say that the sample comes from a population with mean 58 kg. with a variance of 9 kg. ?
11. In a big city two samples of people are drawn. First sample of size 100, the average daily income of people is 210$ with a standard deviation 10$ and in the second sample of size 150 persons, average daily income is 220$ with a standard deviation of 11$. Test if there is any significant difference in average incomes.

### Answers

1. Biased
3. $z = 0.37$ ∴ the difference is not significant
4. $z = 6.56$ ∴ the difference is highly significant
5. $z = 2.8$ ∴ it cannot be regarded as a random sample.
6. Yes
7. No
8. $z = 2.67$ ∴ The claim is not acceptable
9. Yes
10. No
11. $|z| = 7.14$ ∴ the difference is highly significant.

*Chapter 6: Tests of Significance for Small Samples*

# Tests of Significance For Small Samples

## 6.1 Introduction

For small samples (size < 30); tests proposed for large samples do not hold good as sampling distribution cannot be assumed to be normal for small samples. This led to search of new approaches to deal with small samples. It should be made rational that the methods and theory applicable to small samples can be used for large samples; but the converse is not true

After hypothesis formulation; choice of test may be sometimes baffling unless specified which test to use. Following suggestions should be kept in mind while choosing test of significance for any hypothesis.

- Size of sample: If size of sample is greater than thirty, use any of the applicable large sample tests or Chi-Square test depending upon the applicability.
- Variance of population: If the population variance is known and the underlying distribution is normal, $z$-test should be used. Also with known variance; if the distribution is not normal, yet for large sample size, z test can be used. But in case of unknown variance, $t$-test should be used.

**Degree of freedom ($v$):** The degrees of freedom are the number of independent quantities that can be assigned to any statistical distribution arbitrarily. Suppose we are required to choose 5 numbers whose sum is 30, then we have choice of only 4 numbers. This suggests that a data of size $n$ has $(n-1)$ degrees of freedom in general. In case of two restrictions, degrees of freedom will be $(n-2)$.

## 6.2 Student's $t$-Distribution

Gosset; who wrote under the pen-name 'Student', derived a theoretical distribution known as Student's $t$ distribution; which is used to test a hypothesis when the sample size is small and the population variance is not known. As $n$ (size of the sample) increases; $t$ distribution tends to normal distribution. It is evident from the graph below that t-distribution has proportionally larger area at its tails than the normal distribution.

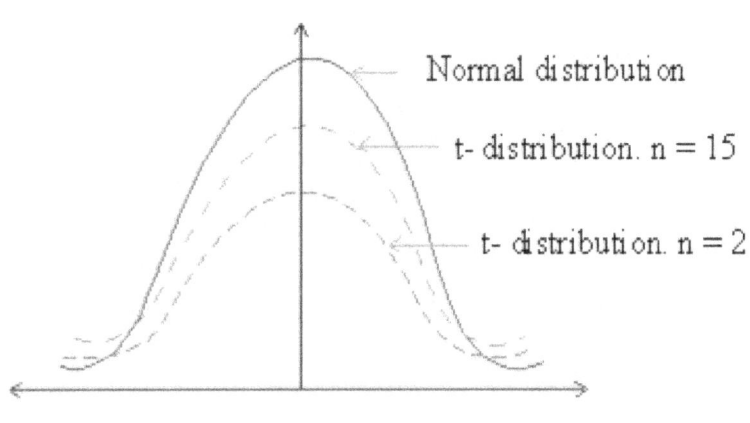

If $x_1, x_2, \cdots, x_n$ be a random sample of size $n \leq 30$, from a normal distribution with mean $\mu$ and variance $\sigma^2$ (not known); also $\bar{x}$ be the sample mean and $s$ standard deviation of the sample

then $t = \dfrac{\bar{x}-\mu}{\frac{s}{\sqrt{n}}}$, where $\bar{x} = \dfrac{\Sigma x}{n}$, $s = \sqrt{\dfrac{\Sigma(x-\bar{x})^2}{n-1}}$

Degree of freedom is given by $\nu = n - 1$

Statistics $t$ is used to test the significance of:
1. Sample mean
2. Difference between two sample means
3. Sample coefficient of correlation
4. Sample Regression Coefficient

**Result 1 Testing significance of sample mean**

Set up the hypothesis $H_0$: sample is drawn from the given population.

To calculate significance of sample mean at 5% level, calculate the statistics

$t = \dfrac{\bar{x}-\mu}{\frac{s}{\sqrt{n}}}$ and compare with the table value of $t$ at $(n-1)$ degrees of freedom.

Let the tabulated value of $t$ be $t_1$, if $|t| < t_1$; $H_0$, is accepted, otherwise value of $t$ is considered significant and we reject the hypothesis $H_0$.

### Fiducial (Confidence) limits of population mean

At 5% significance level (95% confidence level):

$\bar{x} \pm \dfrac{s}{\sqrt{n}} t_{0.05}$, where $t_{0.05}$ is table value of $t$ at given degrees of freedom

At 1% significance level (99% confidence level):

$\bar{x} \pm \dfrac{s}{\sqrt{n}} t_{0.01}$, where $t_{0.01}$ is table value of $t$ at given degrees of freedom.

**Example1** The mean life time of sample of 16 fluorescent light bulbs produced by a company is computed to be 1550 hours with a standard deviation of 100 hours. The company claims that the average life of the each bulb is 1600 hours. Using the level of significance of 0.05, is the claim accepted? Also find the confidence limits for $\mu$.

**Solution**: Let $H_0$: Average life of the each bulb is 1600 hours.

Population mean $(\mu) = 1600$ hours, sample size $(n) = 16$

Sample mean $\bar{x} = 1550$, sample standard deviation $(s) = 100$ hours

Statistic $t = \dfrac{\bar{x}-\mu}{s}\sqrt{n} = \dfrac{1550-1600}{100}(4) = -2$

Table value of $t$ at 5% level of significance for 15 degrees of freedom is 2.13.

$\therefore$ Calculated value of $|t| <$ table value of $t$, hence the hypothesis $(H_0)$ is accepted that the average life of the each bulb is 1600 hours.

Also confidence limits for $\mu = \bar{x} \pm \dfrac{s}{\sqrt{n}} t_{0.05} = 1550 \pm \dfrac{100}{\sqrt{16}}(2.13)$

$= 1496.75$ to $1603.25$

**Example 2** Ten individuals are chosen at random from a population and their heights are found in inches as: 63, 63, 64, 65, 66, 69, 69, 70, 70, and 71. Discuss the suggestion that the mean height of universe is 65. Value of $t$ at 5% level of significance for 9 degrees of freedom is 2.262.

**Solution**: Let $H_0$: The sample is drawn from the given population.

Population mean $(\mu) = 65$, sample size $(n) = 10$

Calculating sample mean and standard deviation:

| $x$ | $x - \bar{x}, \bar{x} = 67$ | $(x - \bar{x})^2$ |
|---|---|---|
| 63 | -4 | 16 |
| 63 | -4 | 16 |
| 64 | -3 | 9 |
| 65 | -2 | 4 |
| 66 | -1 | 1 |
| 69 | +2 | 4 |
| 69 | +2 | 4 |
| 70 | +3 | 9 |
| 70 | +3 | 9 |
| 71 | +4 | 16 |
| $\sum x = 670$ $\bar{x} = \frac{670}{10} = 67$ | | $\sum(x - \bar{x})^2 = 88$ |

Sample standard deviation: $s = \sqrt{\frac{(x-\bar{x})^2}{n-1}} = \sqrt{\frac{88}{9}} = 3.13$

Statistic $t = \frac{\bar{x} - \mu}{s}\sqrt{n} = \frac{(67-65)\sqrt{10}}{3.13} = \frac{2\sqrt{10}}{3.13} = 2.02$

∴ Calculated value of $t$ is less than tabulated value at 9 degrees of freedom.

Hence the hypothesis ($H_0$) is accepted that sample is drawn from the given population i.e. mean height of the universe is 65 inches.

### Result II Testing difference between means of two small samples

Let two independent samples $(x_1, x_2, \cdots, x_{n_1})$; $(y_1, y_2, \cdots, y_{n_2})$ of sizes $n_1$; $n_2$ with means $\bar{x}$; $\bar{y}$ and standard deviations $s_x = \sqrt{\frac{\sum(x-\bar{x})^2}{n_1 - 1}}$; $s_y = \sqrt{\frac{\sum(y-\bar{y})^2}{n_2 - 1}}$ be drawn from two normal populations with means $\mu_1$; $\mu_2$ and having same variance $\sigma^2$, then to test whether the two population means $\mu_1$ and $\mu_2$ are same:

Calculate $t = \frac{\bar{x} - \bar{y}}{s\sqrt{\frac{1}{n_1} + \frac{1}{n_2}}}$; where $s = \sqrt{\frac{\sum(x-\bar{x})^2 + \sum(y-\bar{y})^2}{n_1 + n_2 - 2}}$ or $\sqrt{\frac{(n_1 - 1)s_x^2 + (n_2 - 1)s_y^2}{n_1 + n_2 - 2}}$

and compare with table value of $t$ (say $t_1$) at $(n_1 + n_2 - 2)$ degrees of freedom. Hypothesis is accepted if calculated value of $|t| < t_1$.

**Example 3** Two independent samples showing weights in ounces of eight and seven items are given below:

| Sample I:  | 10 | 12 | 13 | 11 | 15 | 9 | 12 | 14 |
|------------|----|----|----|----|----|---|----|----|
| Sample II: | 9  | 11 | 10 | 13 | 9  | 8 | 10 |    |

Is the difference between the means of the samples significant?

Value of $t$ at 5% level of significance for 15 degrees of freedom is 2.16.

**Solution:** Let the null hypothesis ($H_0$) be: $\mu_1 \approx \mu_2$

Here $n_1 = 8$ and $n_2 = 7$, calculating sample means and standard deviations

| $x$ | $(x - \bar{x})$ | $(x - \bar{x})^2$ | $y$ | $(y - \bar{y})$ | $(y - \bar{y})^2$ |
|-----|-----------------|-------------------|-----|-----------------|-------------------|
| 10  | -2              | 4                 | 9   | -1              | 1                 |
| 12  | 0               | 0                 | 11  | 1               | 1                 |
| 13  | 1               | 1                 | 10  | 0               | 0                 |
| 11  | -1              | 1                 | 13  | 3               | 9                 |
| 15  | 3               | 9                 | 9   | -1              | 1                 |
| 9   | -3              | 9                 | 8   | -2              | 4                 |
| 12  | 0               | 0                 | 10  | 0               | 0                 |
| 14  | 2               | 4                 |     |                 |                   |
| $\sum x = 96$ |        | 28                | $\sum y = 70$ |      | 16                |
| $\bar{x} = \frac{96}{8} = 12$ |  |     | $\bar{y} = \frac{70}{7} = 10$ |  |          |

$$S = \sqrt{\frac{\sum(x-\bar{x})^2 + \sum(y-\bar{y})^2}{n_1 + n_2 - 2}} = \sqrt{\frac{\sum(x-12)^2 + \sum(y-10)^2}{8+7-2}} = \sqrt{\frac{28+16}{8+7-2}} = 1.8$$

$$\therefore t = \frac{\bar{x} - \bar{y}}{S\sqrt{\frac{1}{n_1} + \frac{1}{n_2}}} = \frac{12 - 10}{(1.8)\sqrt{\frac{1}{8} + \frac{1}{7}}} = 2.15$$

Table value of $t$ at 5% level of significance for 13 degrees of freedom is 2.16

$\therefore$ Calculated value of $|t| <$ table value of $t$

Hence the hypothesis ($H_0$) is accepted that $\mu_1 \approx \mu_2$, i.e. difference between the means of the two samples is not significant.

**Example 4** Two types of batteries were tested for their mean life length and following results are obtained. Is there a significant difference in the two batteries?

|           | Sample size | Mean life   | Variance |
|-----------|-------------|-------------|----------|
| Battery A | 10          | 500 hours   | 100      |
| Battery B | 8           | 540 hours   | 81       |

**Solution:** Let the null hypothesis ($H_0$) be: $\mu_1 \approx \mu_2$ i.e. there is no significant difference in the two batteries. Here $n_1 = 10$ and $n_2 = 8$, $\bar{x} = 500$ and $\bar{y} = 540$, $s_x = 10$, $s_y = 9$

$$\Rightarrow S = \sqrt{\frac{(n_1-1)s_x^2 + (n_2-1)s_y^2}{n_1 + n_2 - 2}} = \sqrt{\frac{9(100) + 7(81)}{10+8-2}} = 9.58$$

$$\therefore t = \frac{\bar{x} - \bar{y}}{S\sqrt{\frac{1}{n_1} + \frac{1}{n_2}}} = \frac{500 - 540}{(9.58)\sqrt{\frac{1}{10} + \frac{1}{8}}} = -8.8$$

Table value of $t$ at 5% level of significance for 16 degrees of freedom is 2.12.

∴ Calculated value of $|t| \gg$ table value of $t$, hence the hypothesis $(H_0)$ is rejected that $\mu_1 \approx \mu_2$, i.e. difference between two batteries is highly significant.

**Example5**: A set of 15 observations has mean 68.57; standard deviation 2.4, another set of 7 observations gives mean as 64.14; standard deviation 2.7.

Use $t$-test to find whether two sets of data are drawn from populations with same mean, if standard deviations of two populations are assumed to be equal.

**Solution:** Let the null hypothesis $(H_0)$ be: $\mu_1 \approx \mu_2$

i.e. there is no significant difference between the two population means.

Here $n_1 = 15$ and $n_2 = 7$, $\bar{x} = 68.57$ and $\bar{y} = 64.14$

$s_x = 2.4$, $s_y = 2.7$

$$\Rightarrow s = \sqrt{\frac{(n_1-1)s_x^2 + (n_2-1)s_y^2}{n_1+n_2-2}} = \sqrt{\frac{14(2.4)^2 + 6(2.7)^2}{15+7-2}} = 2.49$$

$$\therefore t = \frac{\bar{x}-\bar{y}}{s\sqrt{\frac{1}{n_1}+\frac{1}{n_2}}} = \frac{68.57-64.14}{(2.49)\sqrt{\frac{1}{15}+\frac{1}{7}}} = 3.89$$

Table value of $t$ at 5% level of significance for 20 degrees of freedom is 2.086

∴ Calculated value of $|t| >$ table value of $t$, hence the hypothesis $(H_0)$ is rejected that $\mu_1 \approx \mu_2$, i.e. difference between two means is significant.

**Result III  Testing difference between two dependent samples or paired observations**

A '$t$' test can be efficiently used to compare two samples of the same population for some treatment effects; for instance to compare efficacy of two drugs on a population or to study the effect of coaching on some students etc.

Compute the statistic $t = \frac{\bar{x}-\Delta}{\frac{s}{\sqrt{n}}}$; where $\Delta$ denotes targeted value and is zero for testing equal means and $s = \sqrt{\frac{\Sigma d^2}{n-1}}$, where $d$ is the deviation from the mean difference.

**Example6** A dietitian opts to try out a new type of diet program on ten overweight girls for 2 months. He targets to make them loose 6 kgs on average, and records their weights before and after the diet program. Use 0.05 significance level to test whether this special diet program helped or not.

| Before weights | 65 | 77 | 99 | 86 | 84 | 93 | 59 | 72 | 69 | 103 |
|---|---|---|---|---|---|---|---|---|---|---|
| After weights | 63 | 72 | 92 | 80 | 80 | 87 | 57 | 67 | 64 | 95 |

Table value of $t$ at 5% level of significance for 9 degrees of freedom is 2.26.

**Solution**: Let $H_0$: Average weight loss caused by diet program is 6 kg.

Calculating deviations from the mean weight loss:

| Girls | Weight difference ($m$) | $d = m - \bar{x}$ | $d^2$ |
|---|---|---|---|
| 1 | -2 | +3 | 9 |
| 2 | -5 | 0 | 0 |
| 3 | -7 | -2 | 4 |
| 4 | -6 | -1 | 1 |
| 5 | -4 | +1 | 1 |
| 6 | -6 | -1 | 1 |
| 7 | -2 | +3 | 9 |
| 8 | -5 | 0 | 0 |
| 9 | -5 | 0 | 0 |
| 10 | -8 | -3 | 9 |
| $n = 10$ | $\sum m = -50$, $\bar{x} = \frac{\sum m}{n} = \frac{-50}{10} = -5$ | | $\sum d^2 = 34$ |

Standard deviation of the differences $s = \sqrt{\frac{\sum d^2}{n-1}} = \sqrt{\frac{34}{9}} = 1.94$

Statistic $t = \frac{\bar{x} - \Delta}{\frac{s}{\sqrt{n}}}$, where targeted weight loss ($\Delta$) = $-6$ kg

$\Rightarrow t = \frac{-5 - (-6)\sqrt{10}}{1.94} = 1.63$ ∴ Calculated value of $|t|$ < table value of $t$

Hence the hypothesis ($H_0$) is accepted that the average weight loss caused by diet program is 6 kg.

**Example 7** Two laboratories carried out independent estimates of lead content (in mg) in noodles of a certain brand. A sample is taken from each batch, halved and the separate halves were tested in the two laboratories to obtain the following results:

| Batch Number | 1 | 2 | 3 | 4 | 5 | 6 | 7 | 8 | 9 | 10 |
|---|---|---|---|---|---|---|---|---|---|---|
| Lab A | 9 | 8 | 8 | 4 | 7 | 7 | 9 | 6 | 6 | 6 |
| Lab B | 7 | 8 | 7 | 3 | 8 | 6 | 9 | 4 | 7 | 8 |

Does the testing suggest same average lead content in the brand?

**Solution:** $H_0$: Average lead content of two samples are equal.

Calculating deviations from the mean difference:

| Batch | Difference ($m$) | $d = m - \bar{x}$ | $d^2$ |
|---|---|---|---|
| 1 | 2 | 1.7 | 2.89 |
| 2 | 0 | -0.3 | 0.09 |
| 3 | 1 | 0.7 | 0.49 |
| 4 | 1 | 0.7 | 0.49 |
| 5 | -1 | -1.3 | 1.69 |
| 6 | 1 | 0.7 | 0.49 |
| 7 | 0 | -0.3 | 0.09 |
| 8 | 2 | 1.7 | 2.89 |
| 9 | -1 | -1.3 | 1.69 |
| 10 | -2 | -2.3 | 5.29 |
| $n = 10$ | $\sum m = 3$, $\bar{x} = \frac{\sum m}{n} = \frac{3}{10} = 0.3$ | | $\sum d^2 = 16.10$ |

Standard deviation of the differences $s = \sqrt{\frac{\Sigma d^2}{n-1}} = \sqrt{\frac{16.1}{9}} = 1.34$

$t = \frac{\bar{x}-\Delta}{\frac{s}{\sqrt{n}}} = \frac{0.3-(0)\sqrt{10}}{1.34} = 0.71$   $\because \Delta = 0$ as we are testing for equal means of two lab tests

Table value of $t$ at 5% level of significance for 9 degrees of freedom is 2.26.

$\therefore$ Calculated value of $|t|$ < table value of $t$ at 9 degrees of freedom at 5% level of significance, hence the hypothesis ($H_0$) is accepted that the average lead content of two samples are equal.

## Result IV Testing significance of population correlation coefficient from sample coefficient of correlation

Let $(x_1, y_1), (x_2, y_2), \ldots, (x_n, y_n)$ having coefficient of correlation $r$, be a random sample from a bivariate frequency distribution having individual means $\mu_1; \mu_2$ and standard deviations $\sigma_1; \sigma_2$. Then to test the hypothesis that population correlation coefficient ($\rho$) is zero: Compute the statistic $t = \frac{r\sqrt{n-2}}{\sqrt{1-r^2}}$, where $n$ is the sample size

Hypothesis is accepted if calculated value of $|t|$ is less than tabulated value of $t$ at

$(n-2)$ degrees of freedom for the specified significance level.

**Example 8** A random sample of size 18 from a bivariate population gave correlation coefficient $r = 0.4$. Does this indicate the existence of correlation in the population?

**Solution**: Let $H_0$: Population correlation coefficient ($\rho$) is zero, i.e. there is no correlation between the population variables.

$$t = \frac{r\sqrt{n-2}}{\sqrt{1-r^2}} = \frac{(0.4)\sqrt{18-2}}{\sqrt{1-(0.4)^2}} = 1.7$$

Table value of $t$ at 16 degrees of freedom is 2.12

$\therefore$ Calculated value of $|t|$ is less than tabulated value of $t$ at 16 degrees of freedom, hence the hypothesis $H_0$ is accepted that there is no correlation between the population variables.

## 6.3 Snedecor's F – test for Testing Equality of Two Population Variances

An $F$-test is used to test if the variances of two populations are equal. It can be a one-tailed or two-tailed test. The one-tailed version tests in only one direction, i.e. variance of the first population is either greater or less than the second population but not both ways. The two-tailed version tests for the hypothesis that the variances are not equal but one can be greater or less than the other.

Let two independent random samples ($x_1, x_2, \cdots, x_{n_1}$); ($y_1, y_2, \cdots, y_{n_2}$) having standard deviations $s_x = \sqrt{\frac{\Sigma(x-\bar{x})^2}{n_1-1}}$; $s_y = \sqrt{\frac{\Sigma(y-\bar{y})^2}{n_2-1}}$ be drawn from two normal populations. Snedecor defined the statistic $F = \frac{s_x^2}{s_y^2}$, $s_x^2 > s_y^2$ for testing equality of two population variances. Here greater of the two variances $s_x^2$ and $s_y^2$ is to be taken in the numerator and if $n_1$ corresponds to the greater variance, then degree of freedom is $(n_1 - 1, n_2 - 1)$.

If calculated value of $F$ is less than the table value of $F$ with $v = (n_1 - 1, n_2 - 1)$ at given level of significance, the null hypothesis that 'the two samples might have been drawn from two normal population with the same variance' is accepted.

**Example 9** Two samples of size 9 and 8 give the sum of squares of deviations from their respective means equal to 160 inches square and 91 inches square respectively. Can they be regarded as drawn from two normal populations with same variances? (Value of $F$ for 8 and 7 degrees of freedom is 3.73)

**Solution**: Let $H_0$: Two samples have been drawn from two normal populations with same variance.

Given $s_x^2 = \frac{\Sigma(x-\bar{x})^2}{n_1-1} = \frac{160}{9-1} = 20$, $s_y^2 = \frac{\Sigma(y-\bar{y})^2}{n_2-1} = \frac{91}{8-1} = 13$

$\therefore F = \frac{s_x^2}{s_y^2} = \frac{20}{13} = 1.54 \quad \because s_x^2 > s_y^2$

Table value of $F$ for $(8, 7)$ degrees of freedom is 3.73, $\therefore$ Calculated value of F is much less than the table value of $F$ at $(8,7)$ degrees of freedom. Hence $H_0$ is accepted, i.e. two samples may be regarded as drawn from two normal populations with same variances.

**Example 10** Show how we can use Student's-$t$ test and Snedecor's $F$ test to decide whether the following two samples have been drawn from the same normal population. Which of the two tests would you apply first and why?

|  | Size | Mean | Sum of squares of deviations from the mean |
|---|---|---|---|
| Sample I | 9 | 68 | 36 |
| Sample II | 10 | 69 | 42 |

Given that $t_{0.5}(17) = 2.11$, $F_{0.5}(9,8) = 3.39$

**Solution**: $H_0$: Two samples have been drawn from the same normal population

To test $H_0$ using Student's-$t$ test, population variance $\sigma^2$ should be same, we can test that two samples have been drawn from two normal populations with same variance using Snedecor's $F$ test. $\therefore$ Snedecor's $F$ test should be applied first.

For the two samples I and II say $(x_1, x_2, \cdots, x_9)$; $(y_1, y_2, \cdots, y_{10})$

$s_x^2 = \frac{\Sigma(x-\bar{x})^2}{n_1-1} = \frac{36}{9-1} = 4.5$, $s_y^2 = \frac{\Sigma(y-\bar{y})^2}{n_2-1} = \frac{42}{10-1} = 4.67$

$\therefore F = \frac{s_y^2}{s_x^2} = \frac{4.67}{4.5} = 1.04 \quad \because s_y^2 > s_x^2$

Table value of $F$ for $(9, 8)$ degrees of freedom is 3.39

$\therefore$ Calculated value of $F$ is much less than the table value of $F$ at $(9, 8)$ degrees of freedom.

Hence the two samples may be regarded as drawn from two normal populations with same variances.

Again to test whether the two population means are same using $t$-test

Calculate $t = \dfrac{\bar{x}-\bar{y}}{s\sqrt{\dfrac{1}{n_1}+\dfrac{1}{n_2}}}$; where $s = \sqrt{\dfrac{(n_1-1)s_x^2 +(n_2-1)s_y^2}{n_1+n_2-2}}$

$\Rightarrow s = \sqrt{\dfrac{(9-1)(4.5)+(10-1)(4.67)}{9+10-2}} = 2.14$

$\therefore t = \dfrac{68-69}{(2.14)\sqrt{\dfrac{1}{9}+\dfrac{1}{10}}} = -1.02$

Table value of $t$ at 5% level of significance for 17 degrees of freedom is 2.11.

$\therefore$ Calculated value of $|t| <$ table value of $t$, hence the hypothesis that two population means are same is accepted.

Thus using Snedecor's $F$ test, we can say that the two samples have been drawn from two normal populations with same variance and also the two population means are same using $t$-test, hence we can conclude that the hypothesis $H_0$ that the two samples have been drawn from the same normal population may be accepted.

## 6.4 Fisher's Z Test for Testing Significance of Correlation Coefficient for Small Samples

If $r$ is the correlation co-efficient of a sample and $\rho$ be the population co-efficient of correlation, then to test the hypothesis that the given sample has been drawn from the population whose coefficient of correlation is $\rho$:

Compute the statistic $\dfrac{z-\xi}{\frac{1}{\sqrt{n-3}}}$, where $z = \dfrac{1}{2}\log_e\left(\dfrac{1+r}{1-r}\right)$ and $\xi = \dfrac{1}{2}\log_e\left(\dfrac{1+\rho}{1-\rho}\right)$

If value of $\left|\dfrac{z-\xi}{\frac{1}{\sqrt{n-3}}}\right| < 1.96$, hypothesis is accepted at 5% level of significance.

**Example11** Test the significance of the correlation $r = 0.6$ for a sample of size 20 against hypothetical population coefficient of correlation $\rho = 0.8$

**Solution**: Let $H_0$: correlation coefficient of population is 0.8

Here $z = \dfrac{1}{2}\log_e\left(\dfrac{1+r}{1-r}\right) = \dfrac{1}{2}\log_e\left(\dfrac{1+0.6}{1-0.6}\right) = 0.693$

$\xi = \dfrac{1}{2}\log_e\left(\dfrac{1+\rho}{1-\rho}\right) = \dfrac{1}{2}\log_e\left(\dfrac{1+0.8}{1-0.8}\right) = 1.099$

Now $\dfrac{z-\xi}{\frac{1}{\sqrt{n-3}}} = \dfrac{0.693-1.099}{\frac{1}{\sqrt{20-3}}} = (0.693 - 1.099)\sqrt{17} = -1.67$

$\therefore \left|\dfrac{z-\xi}{\frac{1}{\sqrt{n-3}}}\right| = 1.67 < 1.96$

Hence the sample may be regarded as coming from population with coefficient of correlation $(\rho) = 0.8$

# Exercise 6

1. A factory makes a machine part with axle diameter of 0.7 inch. A random sample of 10 parts shows a mean diameter of 0.742 inch with a standard deviation of 0.04 inch. On the basis of this sample would you say that the work is inferior?
   Value of $t$ at 5% level of significance for 9 degrees of freedom is 2.262.

2. A random sample of 10 boys had the I.Q. levels: 70, 120, 110, 101, 88, 83, 95, 98, 107 and 100. Does this data support the assumption of population mean I.Q. of 100 at 5% level of significance?

3. In a school the heights of six randomly chosen girls are: 63, 65, 68, 69, 71 and 72 inches and those of nine randomly chosen boys are 61, 62, 65, 66, 69, 70, 71, 72 and 73 inches. Discuss the hypothesis that the girls are taller than boys.
   Value of $t$ at 5% level of significance for 13 degrees of freedom is 1.77.

4. A random sample of size 16 has mean 53. The sum of squares of deviations from the mean is 135. Can the sample be regarded as taken from a population having mean as 56?

5. A new medicine is given to 12 patients whose B.P. increases by 5, 2, 8, -1, 3, 0, -2, 1, 5, 0, 4, 6 units. Can we conclude that the medicine results in increased blood pressure?

6. Two samples of different brands were tested for average life; a sample from first brand of size 7 shows a mean life of 1036 hours with a standard deviation of 40 hours and a sample of size 8 shows a mean life of 1234 hours with a standard deviation of 36 hours. Is the difference in the two sample means significant to conclude that the second brand has more life than first brand?

7. A researcher hypothesizes that people who are allowed to sleep for only four hours will score significantly lower in an objective skills test than people who are allowed to sleep for eight hours. He selects sixteen participants and randomly assigns them to one of two groups. In one group he makes participants sleep for eight hours and in the other group he allows them to sleep only for four hours. The next morning he administers the skill test to all participants. Scores range from 1-9 with high scores representing better performance.

   | Test scores→ | | | | | | | | |
   |---|---|---|---|---|---|---|---|---|
   | 8 hours sleep group | 5 | 7 | 5 | 3 | 5 | 3 | 3 | 9 |
   | 4 hours sleep group | 8 | 1 | 4 | 6 | 6 | 4 | 1 | 2 |

   Test the hypothesis, given that $t_{0.5}(14) = 2.145$

8. A group of 10 rats fed on the diet $A$ and another group of 8 rats fed on the diet $B$ recorded the following increase in weights in a week:

   | | Weight gains (grams)→ | | | | | | | | |
   |---|---|---|---|---|---|---|---|---|---|
   | Diet A | 5 | 6 | 8 | 1 | 12 | 4 | 3 | 9 | 6 | 10 |
   | Diet B | 2 | 3 | 6 | 8 | 10 | 1 | 2 | 8 | - | - |

   Does it show superiority of Diet $A$ over that of Diet $B$?

9. Test runs with 6 models of an experimental engine showed that they operated for 24, 28, 21, 23, 32 and 22 minutes with a gallon of fuel. If the probability of a Type I error is at most 0.01, is this an evidence against the hypothesis that on average this kind of engine will operate for at least 27 minutes per gallon on the same fuel?

10. Test whether the following two samples have been drawn from the same normal population.

|  | Size | Mean | Sum of squares of deviations from the mean |
|---|---|---|---|
| Sample I | 10 | 15 | 90 |
| Sample II | 12 | 14 | 108 |

Given that $t_{0.5}(20) = 2.086$, $F_{0.5}(9,11) = 2.9$

## Answers

1. $t = 3.16$ which is greater than the table value at 5% significance level, ∴ work can be considered to be inferior.
2. $t = 0.62$ ∴ given data supports the mean I.Q. as 100.
3. $t = 0.3031$ ∴ there is no significant difference between the sample means
4. No
5. No
6. $t = 18.15$ ∴ significantly different to conclude that the second brand has more life than first brand
7. $t = 0.847$ ∴ there is no significant difference between the performances of two groups.
8. No
9. No
10. Yes, the two samples can be considered to be drawn from the same normal population.

*Chapter 7: Tests of significance based on Chi-Square*

# Tests of Significance Based on Chi-Square ($\chi^2$)

## 7.1 Chi-Square ($\chi^2$) Test

Tests like $t$, $F$ and $Z$ are based on the assumption that the samples are drawn from a normally distributed population. As these tests require assumptions about the type of population or parameters, these tests are called 'parametric tests'.

Sometimes it is unrealistic to make any rigid assumption about the distribution of the population from which samples are drawn. Studies came out with Chi-Square (read as Ki-square) tests which provide non-parametric approach for testing of goodness of fit and independence of attributes. Although it assumes population to be normal while assessing population variance with sample parameters; thus it is parametric in this case.

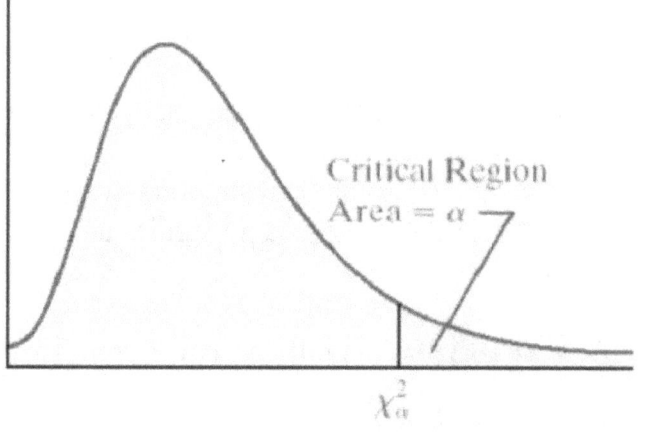

Chi-square ($\chi^2$) is a right tailed test and describes the magnitude of discrepancy between theory and observation. If $\chi^2 = 0$; the observed and expected frequencies completely coincide, $\therefore \chi^2$ provides a measure of correspondence between theory and observations. It is one of the simplest and most general tests and can be used to perform

1. test of significance of sample variance
2. test of goodness of fit
3. test of independence of attributes in a contingency table

## 7.2 Chi-Square ($\chi^2$) Test for Testing Significance of Sample Variance

Chi-Square test can be used to check the population variance for a specified value and also if it needs a revision (if population variance is already given), looking into sample parameters.

If $x_1, x_2, \cdots, x_n$ be a random sample of size $n \leq 30$; with variance $s^2$, from a normal population with given variance $\sigma^2$, then under null hypothesis that population variance is unchanged,

the statistic Chi-Square is defined by: $\chi^2 = \sum \frac{(x-\bar{x})^2}{\sigma^2} = \frac{1}{\sigma^2} \sum (x-\bar{x})^2$ or $\chi^2 = \frac{(n-1)s^2}{\sigma^2}$

where $\bar{x} = \frac{\sum x}{n}$, $s = \sqrt{\frac{\sum(x-\bar{x})^2}{n-1}} \Rightarrow \sum(x-\bar{x})^2 = (n-1)s^2$

$\therefore \chi^2$ is unbiased estimator of testing population variance with degrees of freedom$(\nu) = n-1$.

**Example 1** The standard deviation of a certain dimension of articles produced by a machine is 7.5 over a long period. A random sample of 25 articles gave a standard deviation of 10.0. Is it justified to conclude that variability has increased?

Value of $\chi^2$ for 24 degrees of freedom at 5% significance level is 36.415

**Solution**: Let $H_0$: Population variance $(\sigma^2) = 7.5$ has not changed.

$H_1$: Variance has increased

$$\chi^2 = \frac{(n-1)s^2}{\sigma^2} = \frac{24\,(10)^2}{(7.5)^2} = 42.67$$

Table value of $\chi^2$ for 24 degrees of freedom at 5% level is 36.415

Calculated value of $\chi^2$ is greater than table value. $\therefore$ $H_0$ is rejected and it is justified to conclude that variability has increased.

**Example 2** Eleven measurements of the same object on an instrument; at different times are given by: 2.5, 2.3, 2.4, 2.5, 2.7, 2.5, 2.6, 2.6, 2.7, 2.5 and 2.3.

Test at 1% level of significance that variance of the instrument is not more than 0.16.

**Solution**: Let $H_0$: Population variance $(\sigma^2) = 0.16$

$H_1: \sigma^2 > 0.16$

$$\chi^2 = \frac{1}{\sigma^2}\Sigma(x - \bar{x})^2, \text{ where } \bar{x} = \frac{\Sigma x}{n}$$

Here $\bar{x} = \dfrac{2.5+ 2.3+ 2.4+2.5+2.7+ 2.5+ 2.6+ 2.6+2.7+2.5+ 2.3}{11} = 2.51$

$\therefore \Sigma(x - \bar{x})^2 = 2(2.3 - 2.51)^2 + (2.4 - 2.51)^2 + 4(2.5 - 2.51)^2 +$
$\qquad\qquad\qquad 2(2.6 - 2.51)^2 + 2(2.7 - 2.51)^2 = 0.1891$

$\therefore \chi^2 = \dfrac{1}{\sigma^2}\Sigma(x - \bar{x})^2 = \dfrac{0.1891}{0.16} = 1.182$

Table value of $\chi^2$ for 10 degrees of freedom at 1% level is 23.209

Calculated value of $\chi^2$ is greater than table value. $\therefore$ $H_0$ is rejected and it is justified to conclude that variability has increased.

## 7.3 Chi-Square ($\chi^2$) Test for Testing Goodness of Fit

If $O_i$ $(i = 1,2, ..., n)$ be a set of observed (experimental) frequencies and $E_i$ $(i = 1,2, ..., n)$ be the corresponding set of expected (theoretical) frequencies,

then $\chi^2 = \sum_{i=1}^{n}\dfrac{(O_i - E_i)^2}{E_i}$ ; with degrees of freedom $(v) = n - 1$.

There are some underlying conditions for applying $\chi^2$ test such as:

1. Sum of frequencies should be large (at least 50)
2. No theoretical cell-frequencies should be small, if small (less than 5) theoretical frequencies occur, regrouping of two or more cells must be done before calculating $(O_i - E_i)$. Degree of freedom is determined by the number of classes after regrouping.

**Example 3** Apply $\chi^2$ test of goodness to fit for the following data:

| Observed frequency: | 1 | 5 | 20 | 28 | 42 | 22 | 15 | 5 | 2 |
|---|---|---|---|---|---|---|---|---|---|
| Theoretical Frequency: | 1 | 6 | 18 | 25 | 40 | 25 | 18 | 6 | 1 |

Value of $\chi^2$ for 6, 7, 8 degrees of freedom at 5% significance level are 12.592, 14.067 and 15.507 respectively.

**Solution**: Since first two and the last two cells frequencies are smaller than prescribed, regrouping the data to 7 cells:

| $O_i$ | 6 | 20 | 28 | 42 | 22 | 15 | 7 |
|---|---|---|---|---|---|---|---|
| $E_i$: | 7 | 18 | 25 | 40 | 25 | 18 | 7 |

Degrees of freedom $(v) = 7 - 1 = 6$

$$\chi^2 = \sum_{i=1}^{n} \frac{(O_i - E_i)^2}{E_i}$$

$$= \frac{(6-7)^2}{7} + \frac{(20-18)^2}{18} + \frac{(28-25)^2}{25} + \frac{(42-40)^2}{40} + \frac{(22-25)^2}{25} + \frac{(15-18)^2}{18} + \frac{(7-7)^2}{7} = 1.685$$

Table value of $\chi^2$ for 6 degrees of freedom at 5% level = 12.592

Calculated value of $\chi^2$ is much less than table value ∴ the fit is very good.

**Example4** In experimental breeding, Mendal got the following frequencies of seeds:

315 round and yellow, 101 wrinkled and yellow, 108 round and green, 32 wrinkled and green, total 556. Theory predicts that the frequencies should be proportional in the order 9:3:3:1. Examine the correspondence between theory and experiments.

(Value of $\chi^2$ for 3 degrees of freedom at 5% significance level = 7.815)

**Solution:** Expected (theoretical) frequencies are:

$\frac{9}{16} \times 556 \approx 313, \frac{3}{16} \times 556 \approx 104, \frac{3}{16} \times 556 \approx 104, \frac{1}{16} \times 556 \approx 35$

Comparing observed and expected frequencies:

| $O_i$: | 315 | 101 | 108 | 32 |
|---|---|---|---|---|
| $E_i$ : | 313 | 104 | 104 | 35 |

Degrees of freedom $(v) = 4 - 1 = 3$

$$\chi^2 = \sum_{i=1}^{n} \frac{(O_i - E_i)^2}{E_i}$$

$$= \frac{(315-313)^2}{313} + \frac{(101-104)^2}{104} + \frac{(108-104)^2}{104} + \frac{(32-35)^2}{35} = 0.51$$

Table value of $\chi^2$ for 3 degrees of freedom at 5% level = 7.815

Calculated value of $\chi^2$ is much less than table value, ∴ there is much correspondence between theory and experiment.

**Example5** A die is thrown 246 times and results of these throws are given as:

| Number on the die | 1 | 2 | 3 | 4 | 5 | 6 |
|---|---|---|---|---|---|---|
| Frequency | 32 | 35 | 59 | 57 | 39 | 24 |

Find the value of $\chi^2$ on the hypothesis that the die is unbiased.

**Solution:** Let $H_0$: die is unbiased.

Expected frequency of each number for an unbiased die is $\frac{246}{6} = 41$

Comparing observed and expected frequencies:

| $O_i$ | 32 | 35 | 59 | 57 | 39 | 24 |
|---|---|---|---|---|---|---|
| $E_i:$ | 41 | 41 | 41 | 41 | 41 | 41 |

Degrees of freedom $(\nu) = 6 - 1 = 5$

$$\chi^2 = \sum_{i=1}^{n} \frac{(O_i - E_i)^2}{E_i}$$

$$= \frac{(32-41)^2}{41} + \frac{(35-41)^2}{41} + \frac{(59-41)^2}{41} + \frac{(57-41)^2}{41} + \frac{(39-41)^2}{41} + \frac{(24-41)^2}{41} = 24.15$$

Table value of $\chi^2$ for 5 degrees of freedom at 5% level = 11.07

Calculated value of $\chi^2$ is greater than table value, ∴ the die can be concluded to be biased one.

## 7.4 Chi-Square ($\chi^2$) Test for Testing Independence of Attributes

The Chi-Square test of independence is used to determine if there is a significant relationship between two nominal (categorical) variables.

Consider two attributes $A$ and $B$ to be tested for independence with categories $A_i$, $i = 1, 2, \ldots, m$ and $B_j$, $j = 1, 2, \ldots, n$ respectively.

| Attribute | | $B \rightarrow$ | | | | Total |
|---|---|---|---|---|---|---|
| | | $B_1$ | $B_2$ | | $B_n$ | |
| $A \downarrow$ | $A_1$ | $O_{11}$ | $O_{12}$ | ... | $O_{1n}$ | $\sum_{j=1}^{n} O_{1j}$ |
| | $A_2$ | $O_{21}$ | $O_{22}$ | | $O_{2n}$ | $\sum_{j=1}^{n} O_{2j}$ |
| | ⋮ | | | | | |
| | $A_m$ | $O_{m1}$ | $O_{m2}$ | | $O_{mn}$ | $\sum_{j=1}^{n} O_{mj}$ |
| Total | | $\sum_{i=1}^{m} O_{i1}$ | $\sum_{i=1}^{m} O_{i2}$ | | $\sum_{i=1}^{m} O_{in}$ | $N = \sum_{i,j} O_{ij}$ |

Here $O_{ij}$ is observed frequency for attributes $A_i B_j$ in the given contingency table.

Expected frequency $E_{ij}$ is calculated as shown:

$$E_{11} = \frac{(\sum_{j=1}^{n} O_{1j})(\sum_{i=1}^{m} O_{i1})}{N}, \quad E_{12} = \frac{(\sum_{j=1}^{n} O_{1j})(\sum_{i=1}^{m} O_{i2})}{N}, \ldots, E_{mn} = \frac{(\sum_{j=1}^{n} O_{mj})(\sum_{i=1}^{m} O_{in})}{N}$$

Chi-Square value is given by: $\chi^2 = \sum \frac{(O_{ij} - E_{ij})^2}{E_{ij}}$

For a $m \times n$ contingency table ($m$ rows and $n$ columns), $\nu = (m-1)(n-1)$

**Example 6** Test the hypothesis that flower colour is independent of leaf pattern using the following information:

| Leaves ↓ | Flower colour→ | | |
|---|---|---|---|
| | White | Red | Total |
| Flat | 68 | 32 | 100 |
| Curled | 52 | 58 | 110 |
| Total | 120 | 90 | 210 |

Value of $\chi^2$ for 2 and 1 degree of freedom at 5% level is 5.991 and 3.841 respectively.

**Solution:** Let $H_0$: Flower colour and leaf pattern are independent

On the basis of null hypothesis, expected (theoretical) frequencies are given as:

| Leaves ↓ | Flower colour→ | | |
|---|---|---|---|
| | White | Red | Total |
| Flat | $\frac{(100)(120)}{210} = 57.14$ | $\frac{(100)(90)}{210} = 42.86$ | 100 |
| Curled | $\frac{(110)(120)}{210} = 62.86$ | $\frac{(110)(90)}{210} = 47.14$ | 110 |
| Total | 120 | 90 | 210 |

Degrees of freedom $(v) = (2-1)(2-1) = 1$

$$\chi^2 = \sum \frac{(O_{ij}-E_{ij})^2}{E_{ij}} = \frac{(68-57.14)^2}{57.14} + \frac{(32-42.86)^2}{42.86} + \frac{(52-62.86)^2}{62.86} + \frac{(58-47.14)^2}{47.14} = 9.19$$

Table value of $\chi^2$ for 1 degrees of freedom at 5% level = 3.841

Calculated value of $\chi^2$ is greater than table value, $\therefore H_0$ is rejected, i.e. flower colour and leaf pattern have some connection.

**Example 7** For the attributes $A$ and $B$, in a contingency table $\begin{matrix} a & b \\ c & d \end{matrix}$

show that $\chi^2 = \frac{(a+b+c+d)(ad-bc)^2}{(a+b)(c+d)(b+d)(a+c)}$

**Solution:** The $2 \times 2$ contingency table for the attributes $A$ and $B$ is given below

| Attribute A ↓ | B → | | Total |
|---|---|---|---|
| | $a$ | $b$ | $a+b$ |
| | $c$ | $d$ | $c+d$ |
| Total | $a+c$ | $b+d$ | $N = a+b+c+d$ |

Here $a, b, c, d$ are observed frequencies and under the assumption that the two attributes $A$ and $B$ are independent, expected frequencies are given as $\frac{(a+b)(a+c)}{N}, \frac{(a+b)(b+d)}{N}, \frac{(c+d)(a+c)}{N}, \frac{(c+d)(b+d)}{N}$ respectively.

Now $\chi^2 = \sum \frac{(O_{ij} - E_{ij})^2}{E_{ij}} = \frac{\left[a - \left(\frac{(a+b)(a+c)}{N}\right)\right]^2}{\frac{(a+b)(a+c)}{N}} + \cdots + \frac{\left[d - \left(\frac{(c+d)(b+d)}{N}\right)\right]^2}{\frac{(c+d)(b+d)}{N}}$

$= \frac{[a(a+b+c+d) - (a+b)(a+c)]^2}{(a+b+c+d)(a+b)(a+c)} + \cdots + \frac{[d(a+b+c+d) - (c+d)(b+d)]^2}{(a+b+c+d)(c+d)(b+d)}$

$= \frac{(ad-bc)^2}{(a+b+c+d)} \left[\frac{1}{(a+b)(a+c)}\right] + \cdots + \frac{(ad-bc)^2}{(a+b+c+d)} \left[\frac{1}{(c+d)(b+d)}\right]$

$= \frac{(ad-bc)^2}{(a+b+c+d)} \left[\frac{1}{(a+b)(a+c)} + \frac{1}{(a+b)(b+d)} + \frac{1}{(c+d)(a+c)} + \frac{1}{(c+d)(b+d)}\right]$

$= \frac{(ad-bc)^2}{(a+b+c+d)} \left[\frac{b+d+a+c}{(a+b)(a+c)(b+d)} + \frac{b+d+a+c}{(c+d)(a+c)(b+d)}\right]$

$= \frac{(ad-bc)^2 (a+b+c+d)}{(a+b+c+d)} \left[\frac{c+d+a+b}{(a+b)(a+c)(b+d)(c+d)}\right]$

$= \frac{(a+b+c+d)(ad-bc)^2}{(a+b)(c+d)(b+d)(a+c)}$

**Example 8** A public opinion poll surveyed a simple random sample of 1000 voters. Respondents were classified by voting preferences to parties $A$, $B$ and $C$ and by gender (male or female). Results are shown in the contingency table below.

|  | Voting Preferences | | | Total |
| --- | --- | --- | --- | --- |
|  | Party A | Party B | Party C |  |
| Male | 200 | 150 | 50 | 400 |
| Female | 250 | 300 | 50 | 600 |
| Total | 450 | 450 | 100 | 1000 |

Do the men's voting preferences differ significantly from the women's preferences indicating a gender bias? Use a 0.05 level of significance.

**Solution**: Let $H_0$: Voting preferences and gender are independent.

|  | Voting Preferences | | | Total |
| --- | --- | --- | --- | --- |
|  | Party A | Party B | Party C |  |
| Male | $\frac{(400)(450)}{1000} = 180$ | $\frac{(400)(450)}{1000} = 180$ | $\frac{(400)(100)}{1000} = 40$ | 400 |
| Female | $\frac{(600)(450)}{1000} = 270$ | $\frac{(600)(450)}{1000} = 270$ | $\frac{(600)(100)}{1000} = 60$ | 600 |
| Total | 450 | 450 | 100 | 1000 |

On the basis of null hypothesis, expected frequencies are given in the above table.

Degrees of freedom $(\nu) = (2-1)(3-1) = 2$

$$\chi^2 = \sum \frac{(O_{ij}-E_{ij})^2}{E_{ij}} = \frac{(200-180)^2}{180} + \frac{(150-180)^2}{180} + \frac{(50-40)^2}{40} + \frac{(250-270)^2}{270} + \frac{(300-270)^2}{270} + \frac{(50-60)^2}{60}$$

= 16.20, table value of $\chi^2$ for 2 degrees of freedom at 5% level = 5.991

Calculated value of $\chi^2$ is greater than table value, $\therefore H_0$ is rejected, i.e. voting preferences and gender cannot be considered independent.

## Exercise 7

1. Following tables shows number of male and female births in 800 families having four children:

    | Number of male births | 0 | 1 | 2 | 3 | 4 |
    |---|---|---|---|---|---|
    | Number of female births | 4 | 3 | 2 | 1 | 0 |
    | Number of families | 32 | 178 | 290 | 236 | 94 |

    Test if the data is consistent with the hypothesis that the binomial distribution holds and the probability of mail birth is same as that of a female birth

2. A random sample of 395 people was surveyed and each person was asked to report the highest education level they obtained. The data that resulted from the survey is summarized in the following table:

    |  | Senior Secondary School | Bachelors | Masters | Ph.d. | Total |
    |---|---|---|---|---|---|
    | Female | 60 | 54 | 46 | 41 | 201 |
    | Male | 40 | 44 | 53 | 57 | 194 |
    | Total | 100 | 98 | 99 | 98 | 395 |

Test whether gender and education level independent at 5% level of significance? Value of $\chi^2$ at 3 degrees of freedom is 7.815.

3. Following data shows number of good and bad parts produced by each of the three shifts in a factory.

    |  | Good parts | Bad parts | Total |
    |---|---|---|---|
    | Day shift | 960 | 40 | 1000 |
    | Evening shift | 940 | 50 | 990 |
    | Night shift | 950 | 45 | 995 |
    | Total | 2850 | 135 | 2985 |

Test whether the production of bad parts is independent of the shift on which they were produced. Value of $\chi^2$ at 2 degrees of freedom is 5.991.

4. Out of a sample of 120 persons in a village, 76 persons were administered a new drug for controlling influenza and out of them, 24 persons were attacked by influenza. Out of those who were not administered the new drug, 12 persons were not affected by influenza. Use Chi-square test for finding whether the new drug is effective or not. Value of $\chi^2$ for 1 degree of freedom is 3.84.

5. In a survey of 200 boys of which 75 are intelligent, 40 have educated fathers, while 85 of the unintelligent boys have uneducated fathers. Do these figures support the hypothesis that educated fathers have intelligent boys? Value of $\chi^2$ for 1 degree of freedom is 3.84.

6. The following table shows the number of people having T.B. out of number of enquired people in different age-groups.

| Age-group | Number of people enquired | T.B. cases |
|---|---|---|
| 15-25 | 199 | 1 |
| 25-35 | 300 | 8 |
| 35-45 | 1128 | 38 |
| 45-55 | 1375 | 96 |
| 55-65 | 1089 | 105 |
| 65-75 | 625 | 56 |
| 75-85 | 155 | 12 |
| Total | 4871 | 316 |

Do these figures support the hypothesis that T. B. is equally spread in all age-groups?

Value of $\chi^2$ at 6 degrees of freedom is 12.59

7. Two hundred digits were chosen at random from a set of tables and their frequencies are given below:

| Digits | 0 | 1 | 2 | 3 | 4 | 5 | 6 | 7 | 8 | 9 |
|---|---|---|---|---|---|---|---|---|---|---|
| Frequency | 18 | 19 | 23 | 21 | 16 | 25 | 21 | 20 | 21 | 15 |

Test the hypothesis that the digits were distributed in an equal manner in the tables from which they are chosen. Value of $\chi^2$ at 9 degrees of freedom is 16.919.

Following data gives number of accidents in a city from years 2005 to 2014. Use $\chi^2$ test of goodness of fit to prove the hypothesis that the number of accidents reported for each year from 2005 to 2014 does not differ significantly from an equal number of accidents in each year.

| Year | 2005 | 2006 | 2007 | 2008 | 2009 | 2010 | 2011 | 2012 | 2013 | 2014 |
|---|---|---|---|---|---|---|---|---|---|---|
| Number of Accidents | 164 | 142 | 153 | 171 | 155 | 148 | 136 | 133 | 138 | 140 |

## Answers

1. $\chi^2 = 54.43$ ∴ probability of mail birth is not same as that of a female birth
2. $\chi^2 = 8.006$, which is greater than table value ∴ gender and education levels seem dependent on each other.
3. $\chi^2 = 1.28$, which is less than table value ∴ production of bad parts is independent of the shifts
4. $\chi^2 = 18.97$, which is much greater than table value ∴ null hypothesis is not supported, i.e. the new drug is definitely effective for controlling influenza.
5. $\chi^2 = 8.88$, ∴ education of fathers has a significant effect on intelligence of boys.
6. $\chi^2 = 57.6$, which is much greater than table value ∴ hypothesis is not supported.
7. $\chi^2 = 4.3$, which is less than table value ∴ hypothesis is accepted.

*Chapter 8: Analysis of Variance*

# Analysis of Variance (ANOVA)

## 8.1 Introduction

Introduced by Ronald Fisher; Analysis of Variance (ANOVA) has found wide range of applications in number of statistical measures and it's most useful application is in analyzing the differences among three or more group means. ANOVA is conceptually similar to multiple two-sample $t$-tests, but it is more meticulous as it fits a wide range of practical problems.

## 8.2 One-Way ANOVA

One way ANOVA is a hypothesis test; having only one independent variable and is used to test the equality of three of more population means simultaneously.

Suppose $n$ measurements (or observations) are obtained for each of the $m$ groups being tested for a treatment (or attribute) as shown in table below:

| Groups↓ | Measurements → | | | | | |
|---|---|---|---|---|---|---|
| $X_1$ | $x_{11}$ | $x_{12}$ | ... | $x_{1n}$ | $\sum X_1 = \sum_j x_{1j}$ | $\bar{X}_1 = \frac{\sum X_1}{n} = \frac{1}{n}\sum_j x_{1j}$ |
| $X_2$ | $x_{21}$ | $x_{22}$ | ... | $x_{2n}$ | $\sum X_2 = \sum_j x_{2j}$ | $\bar{X}_2 = \frac{\sum X_2}{n} = \frac{1}{n}\sum_j x_{2j}$ |
| | ⋮ | ⋮ | ⋮ | ⋮ | ⋮ | |
| $X_m$ | $x_{m1}$ | $x_{m2}$ | ... | $x_{mn}$ | $\sum X_m = \sum_j x_{mj}$ | $\bar{X}_m = \frac{\sum X_m}{n} = \frac{1}{n}\sum_j x_{mj}$ |

If $\bar{X}_1, \bar{X}_2, ..., \bar{X}_m$ represent mean of each of the $m$ groups respectively,

then total mean of all measurements $(\bar{X}) = \frac{1}{m}(\bar{X}_1 + \bar{X}_2 + \cdots + \bar{X}_m)$ ... ①

$$= \frac{1}{mn}\left(\sum_j x_{1j} + \sum_j x_{2j} + \cdots + \sum_j x_{mj}\right)$$

$$= \frac{1}{mn}\left(\sum_{ij} x_{ij}\right)$$

If $N = mn$ are total number of observations, then total mean of all measurements is given by

$$\bar{X} = \frac{1}{N}\sum_{ij} x_{ij} \qquad \text{... ②}$$

Now total variation of all measurements $(V_{total}) = \sum_{ij}(x_{ij} - \bar{X})^2$

$\Rightarrow V_{total} = \sum_{ij}[x_{ij}^2 - 2x_{ij}\bar{X} + \bar{X}^2]$

$\qquad = \sum_{ij} x_{ij}^2 - 2\bar{X}\sum_{ij} x_{ij} + N\bar{X}^2 \qquad \because \sum_{ij}\bar{X}^2 = N\bar{X}^2$

$$= \sum_{ij} x_{ij}^2 - 2N\bar{X}^2 + N\bar{X}^2 \qquad \because \sum_{ij} x_{ij} = N\bar{X} \qquad \text{from } ②$$

$$= \sum_{ij} x_{ij}^2 - N\bar{X}^2$$

$$= \sum_{ij} x_{ij}^2 - N\left(\frac{1}{N}\sum_{ij} x_{ij}\right)^2 \qquad \text{from } ②$$

$$= \sum_{ij} x_{ij}^2 - \frac{1}{N}(\sum_i X_i)^2$$

By adding squares of measurements group-wise, we get

$$\therefore V_{total} = \sum X_1^2 + \sum X_2^2 + \cdots + \sum X_m^2 - \frac{(\sum X_1 + \sum X_2 + \cdots + \sum X_m)^2}{N}$$

Again variation between groups $(V_{between}) = n\sum_i (\bar{X}_i - \bar{X})^2$

$$\Rightarrow V_{between} = n\sum_i (\bar{X}_i^2 - 2\bar{X}_i\bar{X} + \bar{X}^2)$$

$$= n[\sum_i \bar{X}_i^2 - 2\bar{X}\sum_i \bar{X}_i + m\bar{X}^2] \qquad \because \sum_i \bar{X}^2 = m\bar{X}^2$$

$$= n[\sum_i \bar{X}_i^2 - 2m\bar{X}^2 + m\bar{X}^2] \qquad \because \sum_i \bar{X}_i = m\bar{X} \quad \text{from } ①$$

$$= n[\sum_i \bar{X}_i^2 - m\bar{X}^2]$$

$$= n\sum_i \left(\frac{1}{n}\sum_j x_{ij}\right)^2 - mn\bar{X}^2 \qquad \because \bar{X}_i = \frac{1}{n}\sum_j x_{ij}$$

$$= \frac{1}{n}\sum_i (\sum_j x_{ij})^2 - N\left(\frac{1}{N}\sum_{ij} x_{ij}\right)^2 \qquad \because mn = N$$

$$= \frac{1}{n}\sum_i (\sum X_i)^2 - \frac{1}{N}(\sum_i X_i)^2 \qquad \because \sum_j x_{ij} = \sum X_i$$

By adding squares of measurements group-wise, we get

$$V_{between} = \frac{(\sum X_1)^2}{n} + \frac{(\sum X_2)^2}{n} + \cdots + \frac{(\sum X_m)^2}{n} - \frac{(\sum X_1 + \sum X_2 + \cdots + \sum X_m)^2}{N}$$

Also $V_{total} = V_{within} + V_{between}$ (where $V_{within}$ gives variation within groups)

$$\therefore V_{within} = V_{total} - V_{between}$$

**Working methodology to compute one-way ANOVA**

Let there be $n$ measurements (or observations) for each of the $m$ groups being tested for a treatment (or attribute). Thus in total there are $N = mn$ measurements.

**Step1**: Assume null hypothesis ($H_0$) that there is no significant difference among the groups being tested for the treatment (or attribute). Then the alternative hypothesis would be that there is at least one significant difference among the groups.

**Step2**: Calculate sum of measurements (observations) for each group as $\sum X_1, \sum X_2, \ldots, \sum X_m$ and compute their squares as $(\sum X_1)^2, (\sum X_2)^2, \ldots, (\sum X_m)^2$

**Step3**: Find square of each element in all the groups and record the sum of squares as $\sum X_1^2, \sum X_2^2, \ldots, \sum X_m^2$.

**Step4**: Now total variation of all measurements is given by:

$$V_{total} = \sum X_1^2 + \sum X_2^2 + \cdots + \sum X_m^2 - \frac{(\sum X_1 + \sum X_2 + \cdots + \sum X_m)^2}{N}$$

**Step5**: Also variation between groups is given by:

$$V_{between} = \frac{(\Sigma X_1)^2}{n} + \frac{(\Sigma X_2)^2}{n} + \cdots + \frac{(\Sigma X_m)^2}{n} - \frac{(\Sigma X_1 + \Sigma X_2 + \cdots + \Sigma X_m)^2}{N}$$

**Step6**: Now compute variations within groups as:

$$V_{within} = V_{total} - V_{between}$$

**Step7**: Compute mean variations as $M_{between} = \frac{V_{between}}{v_1}$ and $M_{within} = \frac{V_{within}}{v_2}$,

Where $v_1 = m - 1$ are the degrees of freedom between groups

$v_2 = N - m$ are the degrees of freedom within groups

**Step8**: Calculate $F = \frac{M_{between}}{M_{within}}$ and compare with table value of $F$ statistic.

Hypothesis ($H_0$) is accepted if calculated value of $|F|$ is less than table value of $F$ at $(v_1, v_2)$ degrees of freedom at the specified level of significance.

➤ If sizes of each group differ say $n_1, n_2, \ldots, n_m$ then $N = n_1 + n_2 + \cdots + n_m$ and also variation between groups is given by:

$$V_{between} = \frac{(\Sigma X_1)^2}{n_1} + \frac{(\Sigma X_2)^2}{n_2} + \cdots + \frac{(\Sigma X_m)^2}{n_m} - \frac{(\Sigma X_1 + \Sigma X_2 + \cdots + \Sigma X_m)^2}{N}$$

**Example1** Three varieties $A, B, C$ of rice are sown in eight plots of equal area and following quantities per acre were recorded:

| Varieties ↓ | Plots → | | | | | | | |
|---|---|---|---|---|---|---|---|---|
| A | 7 | 4 | 6 | 8 | 6 | 6 | 2 | 9 |
| B | 5 | 5 | 3 | 4 | 4 | 7 | 2 | 2 |
| C | 2 | 4 | 7 | 1 | 2 | 1 | 5 | 5 |

Test the significance of difference between the yields of varieties.

**Solution**: Let $H_0$: Rice varieties do not significantly differ from each other.

Number of varieties ($m$) = 3, Plots for each variety ($n$) = 8, ∴ $N = mn = 24$

| Varieties ↓ | | Plots → | | | | | | | | |
|---|---|---|---|---|---|---|---|---|---|---|
| A | $X_1$ | 7 | 4 | 6 | 8 | 6 | 6 | 2 | 9 | $\Sigma X_1 = 48$, $(\Sigma X_1)^2 = 2304$ |
| | $X_1^2$ | 49 | 16 | 36 | 64 | 36 | 36 | 4 | 81 | $\Sigma X_1^2 = 322$ |
| B | $X_2$ | 5 | 5 | 3 | 4 | 4 | 7 | 2 | 2 | $\Sigma X_2 = 32$, $(\Sigma X_2)^2 = 1024$ |
| | $X_2^2$ | 25 | 25 | 9 | 16 | 16 | 49 | 4 | 4 | $\Sigma x_2^2 = 148$ |
| C | $X_3$ | 2 | 4 | 7 | 1 | 2 | 1 | 5 | 5 | $\Sigma X_3 = 27$, $(\Sigma X_3)^2 = 729$ |
| | $X_3^2$ | 4 | 16 | 49 | 1 | 4 | 1 | 25 | 25 | $\Sigma X_3^2 = 125$ |

Calculating total variation ($V_{total}$)

$$V_{total} = \sum X_1^2 + \sum X_2^2 + \sum X_3^2 - \frac{(\sum X_1 + \sum X_2 + \sum X_3)^2}{N}$$

$$\Rightarrow V_{total} = 322 + 148 + 125 - \frac{(48+32+27)^2}{24} = 117.96$$

Now calculating variation between groups ($V_{between}$)

$$V_{between} = \frac{(\sum X_1)^2}{n} + \frac{(\sum X_2)^2}{n} + \frac{(\sum X_3)^2}{n} - \frac{(\sum X_1 + \sum X_2 + \sum X_3)^2}{N}$$

$$\Rightarrow V_{between} = \frac{2304}{8} + \frac{1024}{8} + \frac{729}{8} - \frac{(48+32+27)^2}{24} = 30.08$$

$$\therefore V_{within} = V_{total} - V_{between} = 117.96 - 30.08 = 87.88$$

| Source of Variation | Variation | DF | Mean variations | F |
|---|---|---|---|---|
| Between groups | 30.8 | $v_1 = 2$ | $M_{between} = \frac{30.08}{2} = 15.04$ | $\frac{15.04}{4.18} = 3.59$ |
| Within groups | 87.88 | $v_2 = 21$ | $M_{within} = \frac{87.88}{21} = 4.18$ | |

Table value of $F$ for (2, 21) degrees of freedom at 5% level is 3.47 and at 1% level it is 5.78. $\therefore$ Calculated value of $F$ is less than table value at 1% level but greater than one at 5% level. Thus the hypothesis $H_0$ saying 'Rice varieties do not significantly differ from each other' may be accepted at 1% level of significance but has to be rejected at 5% level.

**Example2** Following table shows monthly income in million dollars of top four firms in each of the three cities.

| Cities↓ | Annual income (million$)→ | | | |
|---|---|---|---|---|
| A | 50 | 49 | 49 | 48 |
| B | 49 | 48 | 48 | 47 |
| C | 51 | 50 | 50 | 49 |

Calculate combined mean monthly trade income of top firms in three cities. Also test the significance of difference in the three cities in terms of monthly trade income of top firms, given that tabulated value of $F$ at (2, 9) degrees of freedom is 4.26.

**Solution**: $H_0$: Three cities do not significantly differ from each other in terms of monthly trade income of top firms.

Subtracting 45 from each income figure to simplify the calculations, we get

| Cities↓ | Annual income (million$)→ | | | |
|---|---|---|---|---|
| A | 5 | 4 | 4 | 3 |
| B | 4 | 3 | 3 | 2 |
| C | 6 | 5 | 5 | 4 |

Number of groups ($m$) = 3, measures for each group ($n$) = 4, $\therefore N = mn = 12$

| Cities↓ | Annual income (million$)→ | | | | | |
|---|---|---|---|---|---|---|
| A | $X_1$ | 5 | 4 | 4 | 3 | $\sum X_1 = 16$, $(\sum X_1)^2 = 256$ |
| | $X_1^2$ | 25 | 16 | 16 | 9 | $\sum X_1^2 = 66$ |
| B | $X_2$ | 4 | 3 | 3 | 2 | $\sum X_2 = 12$, $(\sum X_2)^2 = 144$ |
| | $X_2^2$ | 16 | 9 | 9 | 4 | $\sum X_2^2 = 38$ |
| C | $X_3$ | 6 | 5 | 5 | 4 | $\sum X_3 = 20$, $(\sum X_3)^2 = 400$ |
| | $X_3^2$ | 36 | 25 | 25 | 16 | $\sum X_3^2 = 102$ |

Combined mean for all the income figures $(\bar{X}) = \frac{1}{N}(\sum X_1 + \sum X_2 + \sum X_3)$

$$= \frac{1}{12}(16 + 12 + 20) = 4$$

∴ Actual mean combined monthly trade income of top firms in three cities

$$= 45 + 4 = 49 \text{ million dollars}$$

Calculating total variation $(V_{total})$

$$V_{total} = \sum X_1^2 + \sum X_2^2 + \sum X_3^2 - \frac{(\sum X_1 + \sum X_2 + \sum X_3)^2}{N}$$

$$\Rightarrow V_{total} = 66 + 38 + 102 - \frac{(16+12+20)^2}{12} = 14$$

Now calculating variation between groups $(V_{between})$

$$V_{between} = \frac{(\sum X_1)^2}{n} + \frac{(\sum X_2)^2}{n} + \frac{(\sum X_3)^2}{n} - \frac{(\sum X_1 + \sum X_2 + \sum X_3)^2}{N}$$

$$\Rightarrow V_{between} = \frac{256}{4} + \frac{144}{4} + \frac{400}{4} - \frac{(16+12+20)^2}{12} = 8$$

∴ $V_{within} = V_{total} - V_{between} = 14 - 8 = 6$

| Source of Variation | Variation | DF | Mean variations | F |
|---|---|---|---|---|
| Between groups | 8 | $v_1 = 2$ | $M_{between} = \frac{8}{2} = 4$ | $\frac{4}{0.67} = 5.97$ |
| Within groups | 6 | $v_2 = 9$ | $M_{within} = \frac{6}{9} = 0.67$ | |

Table value of **F** for (2, 9) degrees of freedom at 5% level is 4.26

∴ Calculated value of $F$ is more than table value and thus the hypothesis $H_0$ is rejected at 5% level of significance and hence we can say that the three cities significantly differ from each other in terms of annual trade income of top firms.

**Example 3** A music teacher assumes that students will learn most effectively with constant background music, compared to an unpredictable sound or no sound at all. He randomly divides twelve students into three groups of four each. All students study a passage of text for half an hour. Those in group I study with a sound at a constant volume in the background. Those in group II study with noise that changes volume periodically. Those in group III study with no sound in the background at all. After studying, all students take a multiple choice test of 10 marks over the material and their scores are given in the following table:

| Groups↓ | Test scores→ | | | |
|---|---|---|---|---|
| I | 8 | 4 | 6 | 7 |
| II | 7 | 5 | 5 | 3 |
| III | 2 | 5 | 4 | 4 |

Test the significance of difference of scores in the three groups, given that tabulated value of $F$ at $(2, 9)$ degrees of freedom is 4.26.

**Solution**: $H_0$: test scores of three groups do not significantly differ from each other.

Number of groups $(m) = 3$, Test scores for each group $(n) = 4$, $\therefore N = mn = 12$

| Groups↓ | Test scores→ | | | | | |
|---|---|---|---|---|---|---|
| I | $X_1$ | 8 | 4 | 6 | 7 | $\sum X_1 = 25$, $(\sum X_1)^2 = 625$ |
| | $X_1^2$ | 64 | 16 | 36 | 49 | $\sum X_1^2 = 165$ |
| II | $X_2$ | 7 | 5 | 5 | 3 | $\sum X_2 = 20$, $(\sum X_2)^2 = 400$ |
| | $X_2^2$ | 49 | 25 | 25 | 9 | $\sum X_2^2 = 108$ |
| III | $X_3$ | 2 | 5 | 4 | 4 | $\sum X_3 = 15$, $(\sum X_3)^2 = 225$ |
| | $X_3^2$ | 4 | 25 | 16 | 16 | $\sum X_3^2 = 61$ |

Calculating total variation ($V_{total}$)

$$V_{total} = \sum X_1^2 + \sum X_2^2 + \sum X_3^2 - \frac{(\sum X_1 + \sum X_2 + \sum X_3)^2}{N}$$

$$\Rightarrow V_{total} = 165 + 108 + 61 - \frac{(25+20+15)^2}{12} = 34$$

Now calculating variation between groups ($V_{between}$)

$$V_{between} = \frac{(\sum X_1)^2}{n} + \frac{(\sum X_2)^2}{n} + \frac{(\sum X_3)^2}{n} - \frac{(\sum X_1 + \sum X_2 + \sum X_3)^2}{N}$$

$$\Rightarrow V_{between} = \frac{625}{4} + \frac{400}{4} + \frac{225}{4} - \frac{(25+20+15)^2}{12} = 12.5$$

$\therefore V_{within} = V_{total} - V_{between} = 34 - 12.5 = 21.5$

| Source of Variation | Variation | DF | Mean variations | F |
|---|---|---|---|---|
| Between groups | 12.5 | $v_1 = 2$ | $M_{between} = \frac{12.5}{2} = 6.25$ | $\frac{6.25}{2.39} = 2.62$ |
| Within groups | 21.5 | $v_2 = 9$ | $M_{within} = \frac{21.5}{9} = 2.39$ | |

Table value of $F$ for $(2, 9)$ degrees of freedom at 5% level is 4.26 $\therefore$ Calculated value of $F$ is less than table value and hence the hypothesis $H_0$ saying 'test scores of three groups do not significantly differ from each other' may be accepted at 5% level of significance.

**Example4** Three methods are used in a production procedure and following outputs were obtained:

| Methods↓ | Output → | | | | | | | | | |
|---|---|---|---|---|---|---|---|---|---|---|
| I | 20 | 15 | 18 | 19 | 20 | 22 | 16 | 18 | 29 | 17 |
| II | 18 | 16 | 15 | 22 | 21 | 20 | 18 | | | |
| III | 22 | 20 | 18 | 17 | 18 | 22 | | | | |

Test at 5% level whether the three methods can be considered to be equivalent as far as outputs are concerned.

**Solution**: Let $H_0$: There is no significant difference between the three methods.

Number of methods $(m) = 3$, $n_1 = 10$, $n_2 = 7$, $n_3 = 6$ $\therefore N = n_1 + n_2 + n_3 = 23$

| Variety↓ | | Plots → | | | | | | | | | | |
|---|---|---|---|---|---|---|---|---|---|---|---|---|
| I | $X_1$ | 20 | 15 | 18 | 19 | 20 | 22 | 16 | 18 | 29 | 17 | $\sum X_1 = 194$ $(\sum X_1)^2 = 37636$ |
| | $X_1^2$ | 400 | 225 | 324 | 361 | 400 | 484 | 256 | 324 | 841 | 289 | $\sum X_1^2 = 3904$ |
| II | $X_2$ | 18 | 16 | 15 | 22 | 21 | 20 | 18 | | | | $\sum X_2 = 130$ $(\sum X_2)^2 = 16900$ |
| | $X_2^2$ | 324 | 256 | 225 | 484 | 441 | 400 | 324 | | | | $\sum X_2^2 = 2454$ |
| III | $X_3$ | 22 | 20 | 18 | 17 | 18 | 22 | | | | | $\sum X_3 = 117$ $(\sum X_3)^2 = 13689$ |
| | $X_3^2$ | 484 | 400 | 324 | 289 | 324 | 484 | | | | | $\sum X_3^2 = 2305$ |

Calculating total variation ($V_{total}$)

$$V_{total} = \sum X_1^2 + \sum X_2^2 + \sum X_3^2 - \frac{(\sum X_1 + \sum X_2 + \sum X_3)^2}{N}$$

$$\Rightarrow V_{total} = 3904 + 2454 + 2305 - \frac{(194+130+117)^2}{23} = 207.3$$

Now calculating variation between groups ($V_{between}$)

$$V_{between} = \frac{(\sum X_1)^2}{n_1} + \frac{(\sum X_2)^2}{n_2} + \frac{(\sum X_3)^2}{n_3} - \frac{(\sum X_1 + \sum X_2 + \sum X_3)^2}{N}$$

$$\Rightarrow V_{between} = \frac{37636}{10} + \frac{16900}{7} + \frac{13689}{6} - \frac{(194+130+117)^2}{23} = 3.69$$

$$\therefore V_{within} = V_{total} - V_{between} = 207.3 - 3.69 = 203.61$$

| Source of Variation | Variation | DF | Mean variations | F |
|---|---|---|---|---|
| Between groups | 3.69 | $v_1 = 2$ | $M_{between} = \frac{3.69}{2} = 1.845$ | $\frac{1.845}{10.18} = 0.18$ |
| Within groups | 203.61 | $v_2 = 20$ | $M_{within} = \frac{203.61}{20} = 10.18$ | |

Table value of **F** for (2, 20) degrees of freedom at 5% level is 3.49 ∴ Calculated value of *F* is much less than table value at 5% level Thus the hypothesis $H_0$ saying 'There is no significant difference between the three methods' may be accepted at 5% level of significance.

## 8.3 Two-Way ANOVA

Two ways ANOVA comprising of two independent variables (called factors); is a statistical technique wherein the interaction between factors and influencing variables can be studied.

Two-Way ANOVA is useful when we need to compare the effect of multiple levels of two factors and there are multiple observations at each level.

**Working methodology to compute two-way ANOVA**

Let there be $m$ rows of combination of a given factor; each row subject to $n$ columns for another specified factor (treatment). Thus in total there are $N = mn$ measurements such that each measurement corresponds to two factors (say factor I and factor II) at a time.

Then to test two simultaneous null hypotheses:

$H_{01}$: There is no significant difference in factor I.

$H_{02}$: There is no significant difference in application of factor II.

**Step1**: Calculate sum of measurements (observations) for factor I as $\sum X_1, \sum X_2, ..., \sum X_m$ and compute their squares as $(\sum X_1)^2, (\sum X_2)^2, ..., (\sum X_m)^2$.

**Step2**: Find square of each element row-wise and record the sum of squares as $\sum X_1^2, \sum X_2^2, ..., \sum X_m^2$.

**Step3**: Calculate sum of measurements (observations) for factor II as $\sum Y_1, \sum Y_2, ..., \sum Y_n$ and compute their squares as $(\sum Y_1)^2, (\sum Y_2)^2, ..., (\sum Y_n)^2$.

**Step4**: Find square of each element column-wise and record the sum of squares as $\sum Y_1^2, \sum Y_2^2, ..., \sum Y_n^2$.

**Step5**: Calculate total variation of all measurements given by:

$$V_{total} = \sum X_1^2 + \sum X_2^2 + \cdots + \sum X_m^2 - \frac{(\sum X_1 + \sum X_2 + \cdots + \sum X_m)^2}{N}$$

$$\text{Or } V_{total} = \sum Y_1^2 + \sum Y_2^2 + \cdots + \sum Y_n^2 - \frac{(\sum Y_1 + \sum Y_2 + \cdots + \sum Y_n)^2}{N}$$

**Step 6**: Compute variations between factor I given by:

$$V_{between\ (X)} = \frac{(\sum X_1)^2}{n} + \frac{(\sum X_2)^2}{n} + \cdots + \frac{(\sum X_m)^2}{n} - \frac{(\sum X_1 + \sum X_2 + \cdots + \sum X_m)^2}{N}$$

**Step 7**: Compute variations between factor II given by:

$$V_{between\ (Y)} = \frac{(\sum Y_1)^2}{m} + \frac{(\sum Y_2)^2}{m} + \cdots + \frac{(\sum Y_n)^2}{m} - \frac{(\sum Y_1 + \sum Y_2 + \cdots + \sum Y_n)^2}{N}$$

**Step 8**: Now compute variations within two factors I and II as:

$$V_{within(XY)} = V_{total} - V_{between\ (X)} - V_{between\ (Y)}$$

**Step 9**: Taking $v_1 = m - 1$ as the degrees of freedom for factor I,

$v_2 = n - 1$ as the degrees of freedom for factor II

$v_3 = N - m - n$ as degrees of freedom within two factors I and II,

compute mean variations as $M_{between\ (X)} = \dfrac{V_{between\ (X)}}{v_1}$

$$M_{between\ (Y)} = \frac{V_{between\ (Y)}}{v_2}$$

$$M_{within(XY)} = \frac{V_{within(XY)}}{v_3}$$

**Step 10**: Calculate $F_1 = \dfrac{M_{between\ (X)}}{M_{within(XY)}}$ and compare with table value of $F$ statistic.

Hypothesis $(H_{01})$ is accepted if calculated value of $|F_1|$ is less than table value of $F$ at $(v_1, v_3)$ degrees of freedom at the specified level of significance.

Also compute $F_2 = \dfrac{M_{between\ (Y)}}{M_{within(XY)}}$ and compare with table value of $F$ statistic.

Hypothesis $(H_{02})$ is accepted if calculated value of $|F_2|$ is less than table value of $F$ at $(v_2, v_3)$ degrees of freedom at the specified level of significance.

**Example 5** Twenty four dogs were divided into four groups of six each as per their weights (in kilograms) at the beginning of the experiment, so that factor $I$ (initial weight) has 4 levels $W_1$, $W_2$, $W_3$, $W_4$ and each group member was fed with factor II i.e. different diets $D_1, D_2, \ldots, D_6$. Weight gains at different levels on specific diets are given in the table below:

| Groups↓ | Diet→ | | | | | |
|---|---|---|---|---|---|---|
| | $D_1$ | $D_2$ | $D_3$ | $D_4$ | $D_5$ | $D_6$ |
| $W_1$ | 9 | 12 | 9 | 6 | 9 | 10 |
| $W_2$ | 7 | 3 | 7 | 7 | 5 | 5 |
| $W_3$ | 6 | 5 | 9 | 11 | 3 | 11 |
| $W_4$ | 6 | 8 | 11 | 2 | 2 | 10 |

Test the hypothesis that neither the initial weights nor the type of diet affects the weight gain.

**Solution**: $H_{01}$: There is no difference in weight gains.

$H_{02}$: There is no effect of type of diet.

Number of levels of weights $(m) = 4$, and types of diets $(n) = 6$, $\therefore N = mn = 24$

| Groups↓ | Diet→ | | | | | | Sum $\sum X_i$ ($\sum X_i^2$) | $(\sum X_i)^2$ |
|---|---|---|---|---|---|---|---|---|
| | $D_1$ ($Y_1$) | $D_2$ ($Y_2$) | $D_3$ ($Y_3$) | $D_4$ ($Y_4$) | $D_5$ ($Y_5$) | $D_6$ ($Y_6$) | | |
| $W_1$ ($X_1$) | 9 (81) | 12 (144) | 9 (81) | 6 (36) | 9 (81) | 10 (100) | 55 $\sum X_1^2 = 523$ | 3025 |
| $W_2$ ($X_2$) | 7 (49) | 3 (9) | 7 (49) | 7 (49) | 5 (25) | 5 (25) | 34 $\sum X_2^2 = 206$ | 1156 |
| $W_3$ ($X_3$) | 6 (36) | 5 (25) | 9 (81) | 11 (121) | 3 (9) | 11 (121) | 45 $\sum X_3^2 = 393$ | 2025 |
| $W_4$ ($X_4$) | 6 (36) | 8 (64) | 11 (121) | 2 (4) | 2 (4) | 10 (100) | 39 $\sum X_4^2 = 329$ | 1521 |
| Sum $\sum Y_j$ | 28 | 28 | 36 | 26 | 19 | 36 | | |
| $\sum Y_j^2$ | 202 | 242 | 332 | 210 | 119 | 346 | | |
| $(\sum Y_j)^2$ | 784 | 784 | 1296 | 676 | 361 | 1296 | | |

Here numbers in brackets in each cell are the squares of each observation.

Calculating total variation $(V_{total})$

$$V_{total} = \sum X_1^2 + \sum X_2^2 + \sum X_3^2 + \sum X_4^2 - \frac{(\sum X_1 + \sum X_2 + \sum X_3 + \sum X_4)^2}{N}$$

$$\Rightarrow V_{total} = 523 + 206 + 393 + 329 - \frac{(55+34+45+39)^2}{24} = 203.96$$

Now calculating variation between weight gains (factor I)

$$V_{between\ (X)} = \frac{(\sum X_1)^2}{n} + \frac{(\sum X_2)^2}{n} + \frac{(\sum X_3)^2}{n} + \frac{(\sum X_4)^2}{n} - \frac{(\sum X_1 + \sum X_2 + \sum X_3 + \sum X_4)^2}{N}$$

$$\Rightarrow V_{between\ (X)} = \frac{3025}{6} + \frac{1156}{6} + \frac{2025}{6} + \frac{1521}{6} - \frac{(55+34+45+39)^2}{24} = 40.79$$

Now calculating variation between diets (factor II)

$$V_{between\ (Y)} = \frac{(\sum Y_1)^2}{m} + \frac{(\sum Y_2)^2}{m} + \cdots + \frac{(\sum Y_6)^2}{m} - \frac{(\sum Y_1 + \sum Y_2 + \cdots + \sum Y_6)^2}{N}$$

$$\Rightarrow V_{between\,(Y)} = \frac{2(784)}{4} + \frac{2(1296)}{4} + \frac{676}{4} + \frac{361}{4} - \frac{(2(28)+2(36)+26+19)^2}{24} = 52.21$$

$$\therefore V_{within\,(XY)} = V_{total} - V_{between\,(X)} - V_{between\,(Y)}$$

$$= 203.96 - 40.79 - 52.21 = 110.96$$

| Source of Variation | Variation | DF | Mean variations | F |
|---|---|---|---|---|
| Between groups | 40.79 | $v_1 = 3$ | $M_{between\,(X)} = \frac{40.79}{3} = 13.6$ | $F_1 = \frac{13.6}{7.4} = 1.84$ |
| Between diets | 52.21 | $v_2 = 5$ | $M_{between\,(Y)} = \frac{52.21}{5} = 10.44$ | $F_2 = \frac{10.44}{7.4} = 1.41$ |
| Within groups and diets | 110.96 | $v_3 = 15$ | $M_{within\,(XY)} = \frac{110.96}{15} = 7.4$ | |

Table value of **F** for (2, 15) degrees of freedom at 5% level is 3.29 and for (5, 15) degrees of freedom is 2.90. ∴ Calculated values of *F* are less than table values for both weight gains and diets.

Hence the hypothesis $H_{01}$ saying 'There is no difference in weight gains' may be accepted at 5% level of significance.

Also hypothesis $H_{02}$ saying 'There is no effect of type of diet' may be accepted at 5% level of significance.

### Exercise 8

1. A transport company wishes to test the average life of the four tyre brands it uses. The company uses all the tyre brands on randomly selected trucks. Following data shows tyre lives (in thousand miles):

| Brand I | Brand II | Brand III | Brand IV |
|---|---|---|---|
| 20 | 19 | 21 | 15 |
| 23 | 15 | 19 | 17 |
| 18 | 17 | 20 | 16 |
| 17 | 20 | 17 | 18 |
|  | 16 | 16 |  |

Test the hypothesis that average life of each brand is same at 5% significance level.

2. A professor wishes to check the responses of text books from four authors A, B, C, D in the market. He has 37 students whom he distributes at random into four groups of 9, 10, 11

and 7 students and assigns them with a book at random. After the course is over, all the students are given with a test and their scores are shown in the table given below:

| Text books → | Book A | Book B | Book C | Book D |
|---|---|---|---|---|
| Test Scores | 68 | 41 | 54 | 44 |
| | 68 | 47 | 44 | 51 |
| | 69 | 54 | 51 | 69 |
| | 60 | 65 | 56 | 59 |
| | 73 | 32 | 47 | 59 |
| | 64 | 73 | 61 | 55 |
| | 71 | 44 | 59 | 66 |
| | 67 | 48 | 49 | |
| | 75 | 64 | 41 | |
| | | 54 | 61 | |
| | | | 73 | |

Test the hypothesis that all the books are equally good.

3. The varieties A, B, C, D, E, F, G of rice are sown in 5 plots each and the following yields in quintals per acre were obtained:

| A | B | C | D | E | F | G |
|---|---|---|---|---|---|---|
| 13 | 15 | 14 | 14 | 17 | 15 | 16 |
| 11 | 11 | 10 | 10 | 15 | 9 | 12 |
| 10 | 13 | 12 | 15 | 14 | 13 | 13 |
| 16 | 18 | 13 | 17 | 19 | 14 | 15 |
| 12 | 12 | 11 | 10 | 12 | 10 | 11 |

Do the data indicate the significant difference in yields of the varieties?

4. Following are weekly sales records (in thousand Rs.) of three salesmen A, B and C of a company during 13 sale-calls.

| Salesmen↓ | Sale records→ | | | | |
|---|---|---|---|---|---|
| A | 300 | 400 | 300 | 500 | |
| B | 600 | 300 | 300 | 400 | |
| C | 700 | 300 | 400 | 600 | 500 |

Test whether the three salesmen are different.

5. Data given below represents number of units of production per day turned out by 5 different workmen using different types of machines.

| Workmen ↓ | Machine type → | | | |
|---|---|---|---|---|
| | A | B | C | D |
| 1 | 46 | 40 | 49 | 38 |
| 2 | 48 | 42 | 54 | 45 |
| 3 | 36 | 38 | 46 | 34 |
| 4 | 35 | 40 | 48 | 35 |
| 5 | 40 | 44 | 51 | 41 |

i. Test whether the machine productivity is same for the four brands of machine types.

ii. Test whether the 5 workmen significantly differ with respect to productivity.

6. An experiment was conducted to determine the effects of different months of planting and different methods of planting on the yield of sugarcane. Following data gives the production output of sugarcane in 100kgs.

| Method of planting ↓ | Month of planting → | | | |
|---|---|---|---|---|
| | October | November | March | April |
| I | 7.10 | 3.69 | 4.70 | 1.90 |
| II | 10.29 | 4.79 | 4.58 | 2.64 |
| III | 8.30 | 3.58 | 4.90 | 1.80 |

Carry out analysis of variance of the above data to test the significance of planting months as well as methods of planting.

## Answers

1. $F = 1.67$, ∴There is no significant difference in the average lives of four brands.
2. $F = 5.66$, ∴Hypothesis is rejected, thus the books significantly differ from each other.
3. $F = 1.1$, ∴There is no significant difference in yields of different varieties.
4. $F = 1.072$, ∴There is no significant difference in the sales records of three salesmen.
5. $F$ (Workmen) $= 8.20$, significant, $F$ (Machines) $= 19.20$, significant
   ∴ There is significant difference between the productivities of 4 machines and also the 5 workmen in terms of productivity.
6. $F$ (Methods of planting) $= 2.98$, $F$ (Months of planting) $= 19.20$ significant
   ∴ Methods of planting are not significant different but months of planting have significant effect on yield of sugarcane.

# Appendix

**Scilab code to find mean, S.D. and first *r* moments about mean of a grouped data**

```
//Program to find mean, S.D. and first r moments about mean of given grouped data
clc
n=input('Enter the no. of observations:');
disp('Enter the values of xi');
for i=1:n
    x(i)=input('\');
end;
disp('Enter the corresponding frequencies fi:')
sum=0;
for i=1:n
    f(i)=input('\');
    sum=sum+f(i);
end;
r=input('How many moments to be calculated:');
sum1=0
for i=1:n
    sum1=sum1+f(i)*x(i);
end
A=sum1/sum;    //Calculate the average
printf('Average=%f\n',A);
for j=1:r
    sum2=0;
    for i=1:n
        y(i)=f(i)*(x(i)-A)^j;
        sum2=sum2+y(i);
    end
  M(j)=(sum2/sum);   //Calculate the moments
 printf('Moment about mean M(%d)=%f\n',j,M(j));
end
sd=sqrt(M(2));    //Calculate the standard deviation
printf('Standard deviation=%f\n',sd);
```

**Scilab code to find mean, S.D. and first *r* moments about mean of an ungrouped grouped data**

```
// program to find mean, mode, median, moments, skewness and kurtosis of linear data
clc
```

```
function [ ]=moments(A)
// save file as moments.sce
B=gsort(A);
n = length(B);
meanA = sum(B)/n;
if pmodulo(n,2)==0
medianA =((B(n/2)+B(n/2 +1)))/2;
else medianA = B((n+1)/2);
end
C = diff(B)
//Y= diff(X) calculates differences between adjacent elements of X along the first array dimension whose size does not equal 1:
//If X is a vector of length m, then Y = diff(X) returns a vector of length m-1. The elements of Y are the differences between adjacent elements of X.
        //Y = [X(2)-X(1) X(3)-X(2) ... X(m)-X(m-1)]
D = find(C) //D = find(C) finds the idices(positions), where value is non zero
E = diff(D)
[m k] = max(E) // maximum 'm' at kth position
modeA = B(D(k)+1)
printf('Mean of the given data is : %f \n\n', meanA);
printf('Median of the given data is : %f \n\n', medianA);
printf('Mode of the given data is : %f \n\n', modeA);
printf('First moment about the mean(M1)= %f \n\n', 0);
for i=1:n
X(i)=A(i)-meanA;
 end
M2 = sum(X.*X)/n;
M3 = sum(X.*X.*X)/n;
M4 = sum(X.*X.*X.*X)/n;
printf('Second moment about the mean(M2)= %f \n\n', M2);
printf('Third moment about the mean(M3)= %f \n\n', M3);
printf('Fourth moment about the mean(M4)= %f \n\n', M4);
sd= sqrt (M2);
printf('Standard deviation: %f \n\n', sd);
Csk= (meanA - modeA)/sd;
printf('Coefficient of skewness: %f \n\n', Csk);
Sk= (M3)^2/(M2)^3;
printf('Skewness: %f \n\n', Sk);
```

```
Kur= M4/(M2)^2;
printf('Kurtosis: %f \n\n', Kur);
endfunction
```

### Scilab code to find a line of regression

```
// Regression line
clc
n=input('Enter the number of terms:')
  printf(' Enter the values of xi')
   for i=1:n
   x(i)=input('\');
   end
  printf(' Enter the values of yi')
   for i=1:n
   y(i)=input('\');
   end
     sumx=0;sumy=0;sumxy=0;sumx2=0;
   for i=1:n
    sumx=sumx +x(i);
     sumx2=sumx2 +x(i)*x(i);
     sumy=sumy +y(i);
     sumxy=sumxy +x(i)*y(i);
   end
   a=((sumx2*sumy -sumx*sumxy)*1.0/(n*sumx2-sumx*sumx)*1.0);
   b=((n*sumxy-sumx*sumy)*1.0/(n*sumx2-sumx*sumx)*1.0);
   printf ('The line is Y=%3.3f +%3.3f X',a,b)
```

### Scilab code to fit a straight line ($y = ax + b$) to given set of data points

```
// Program of straight line fitting for given n pairs of values (x,y)
clc;
n=input('Enter the no. of pairs of values (x,y):')
disp('Enter the values of x:')
for i=1:n
   x(i)=input(' ')
end
disp('Enter the corresponding values of y:')
for i=1:n
   y(i)=input(' ')
end
sumx=0;sumx2=0;sumy=0;sumxy=0
for i=1:n
```

```
    sumx=sumx+x(i);
    sumx2=sumx2+x(i)*x(i);
    sumy=sumy+y(i);
    sumxy=sumxy+x(i)*y(i);
end
A=[sumx n; sumx2 sumx];
B=[sumy;sumxy];
C=inv(A)*B
printf ('The fitted line is y = (%g)x+(%g)',C(1,1),C(2,1))
```

## Scilab code to fit a parabola ($y = a + bx + cx^2$) to given set of data points

```
clc;
n=input('Enter the no. of pairs of values (x,y):')
disp('Enter the values of x:')
for i=1:n
    x(i)=input(' ')
end
disp('Enter the corresponding values of y:')
for i=1:n
    y(i)=input(' ')
end
sumx=0;sumx2=0;sumx3=0;sumx4=0;sumy=0;sumxy=0;sumx2y=0;
for i=1:n
    sumx=sumx+x(i);
    sumx2=sumx2+x(i)^2;
    sumx3=sumx3+x(i)^3;
    sumx4=sumx4+x(i)^4;
    sumy=sumy+y(i);
    sumxy=sumxy+x(i)*y(i);
    sumx2y=sumx2y+x(i)^2*y(i);
end
A=[n sumx sumx2; sumx sumx2 sumx3;sumx2 sumx3 sumx4];
B=[sumy;sumxy;sumx2y];
C=inv(A)*B
printf('The fitted parabola is y=(%g)+(%g)x+(%g)x^2',C(1,1),C(2,1),C(3,1))
```

# Table1: Area Under Standard Normal Curve

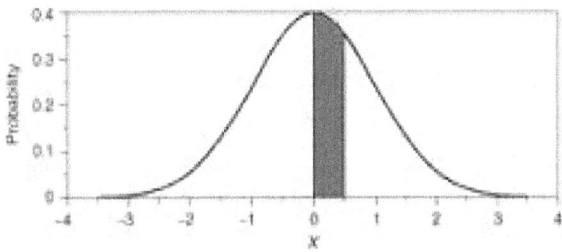

Area under the Normal Curve from 0 to X

| X | 0.00 | 0.01 | 0.02 | 0.03 | 0.04 | 0.05 | 0.06 | 0.07 | 0.08 | 0.09 |
|---|------|------|------|------|------|------|------|------|------|------|
| 0.0 | 0.00000 | 0.00399 | 0.00798 | 0.01197 | 0.01595 | 0.01994 | 0.02392 | 0.02790 | 0.03188 | 0.03586 |
| 0.1 | 0.03983 | 0.04380 | 0.04776 | 0.05172 | 0.05567 | 0.05962 | 0.06356 | 0.06749 | 0.07142 | 0.07535 |
| 0.2 | 0.07926 | 0.08317 | 0.08706 | 0.09095 | 0.09483 | 0.09871 | 0.10257 | 0.10642 | 0.11026 | 0.11409 |
| 0.3 | 0.11791 | 0.12172 | 0.12552 | 0.12930 | 0.13307 | 0.13683 | 0.14058 | 0.14431 | 0.14803 | 0.15173 |
| 0.4 | 0.15542 | 0.15910 | 0.16276 | 0.16640 | 0.17003 | 0.17364 | 0.17724 | 0.18082 | 0.18439 | 0.18793 |
| 0.5 | 0.19146 | 0.19497 | 0.19847 | 0.20194 | 0.20540 | 0.20884 | 0.21226 | 0.21566 | 0.21904 | 0.22240 |
| 0.6 | 0.22575 | 0.22907 | 0.23237 | 0.23565 | 0.23891 | 0.24215 | 0.24537 | 0.24857 | 0.25175 | 0.25490 |
| 0.7 | 0.25804 | 0.26115 | 0.26424 | 0.26730 | 0.27035 | 0.27337 | 0.27637 | 0.27935 | 0.28230 | 0.28524 |
| 0.8 | 0.28814 | 0.29103 | 0.29389 | 0.29673 | 0.29955 | 0.30234 | 0.30511 | 0.30785 | 0.31057 | 0.31327 |
| 0.9 | 0.31594 | 0.31859 | 0.32121 | 0.32381 | 0.32639 | 0.32894 | 0.33147 | 0.33398 | 0.33646 | 0.33891 |
| 1.0 | 0.34134 | 0.34375 | 0.34614 | 0.34849 | 0.35083 | 0.35314 | 0.35543 | 0.35769 | 0.35993 | 0.36214 |
| 1.1 | 0.36433 | 0.36650 | 0.36864 | 0.37076 | 0.37286 | 0.37493 | 0.37698 | 0.37900 | 0.38100 | 0.38298 |
| 1.2 | 0.38493 | 0.38686 | 0.38877 | 0.39065 | 0.39251 | 0.39435 | 0.39617 | 0.39796 | 0.39973 | 0.40147 |
| 1.3 | 0.40320 | 0.40490 | 0.40658 | 0.40824 | 0.40988 | 0.41149 | 0.41308 | 0.41466 | 0.41621 | 0.41774 |
| 1.4 | 0.41924 | 0.42073 | 0.42220 | 0.42364 | 0.42507 | 0.42647 | 0.42785 | 0.42922 | 0.43056 | 0.43189 |
| 1.5 | 0.43319 | 0.43448 | 0.43574 | 0.43699 | 0.43822 | 0.43943 | 0.44062 | 0.44179 | 0.44295 | 0.44408 |
| 1.6 | 0.44520 | 0.44630 | 0.44738 | 0.44845 | 0.44950 | 0.45053 | 0.45154 | 0.45254 | 0.45352 | 0.45449 |
| 1.7 | 0.45543 | 0.45637 | 0.45728 | 0.45818 | 0.45907 | 0.45994 | 0.46080 | 0.46164 | 0.46246 | 0.46327 |
| 1.8 | 0.46407 | 0.46485 | 0.46562 | 0.46638 | 0.46712 | 0.46784 | 0.46856 | 0.46926 | 0.46995 | 0.47062 |
| 1.9 | 0.47128 | 0.47193 | 0.47257 | 0.47320 | 0.47381 | 0.47441 | 0.47500 | 0.47558 | 0.47615 | 0.47670 |
| 2.0 | 0.47725 | 0.47778 | 0.47831 | 0.47882 | 0.47932 | 0.47982 | 0.48030 | 0.48077 | 0.48124 | 0.48169 |
| 2.1 | 0.48214 | 0.48257 | 0.48300 | 0.48341 | 0.48382 | 0.48422 | 0.48461 | 0.48500 | 0.48537 | 0.48574 |
| 2.2 | 0.48610 | 0.48645 | 0.48679 | 0.48713 | 0.48745 | 0.48778 | 0.48809 | 0.48840 | 0.48870 | 0.48899 |
| 2.3 | 0.48928 | 0.48956 | 0.48983 | 0.49010 | 0.49036 | 0.49061 | 0.49086 | 0.49111 | 0.49134 | 0.49158 |
| 2.4 | 0.49180 | 0.49202 | 0.49224 | 0.49245 | 0.49266 | 0.49286 | 0.49305 | 0.49324 | 0.49343 | 0.49361 |
| 2.5 | 0.49379 | 0.49396 | 0.49413 | 0.49430 | 0.49446 | 0.49461 | 0.49477 | 0.49492 | 0.49506 | 0.49520 |
| 2.6 | 0.49534 | 0.49547 | 0.49560 | 0.49573 | 0.49585 | 0.49598 | 0.49609 | 0.49621 | 0.49632 | 0.49643 |
| 2.7 | 0.49653 | 0.49664 | 0.49674 | 0.49683 | 0.49693 | 0.49702 | 0.49711 | 0.49720 | 0.49728 | 0.49736 |
| 2.8 | 0.49744 | 0.49752 | 0.49760 | 0.49767 | 0.49774 | 0.49781 | 0.49788 | 0.49795 | 0.49801 | 0.49807 |
| 2.9 | 0.49813 | 0.49819 | 0.49825 | 0.49831 | 0.49836 | 0.49841 | 0.49846 | 0.49851 | 0.49856 | 0.49861 |
| 3.0 | 0.49865 | 0.49869 | 0.49874 | 0.49878 | 0.49882 | 0.49886 | 0.49889 | 0.49893 | 0.49896 | 0.49900 |
| 3.1 | 0.49903 | 0.49906 | 0.49910 | 0.49913 | 0.49916 | 0.49918 | 0.49921 | 0.49924 | 0.49926 | 0.49929 |
| 3.2 | 0.49931 | 0.49934 | 0.49936 | 0.49938 | 0.49940 | 0.49942 | 0.49944 | 0.49946 | 0.49948 | 0.49950 |
| 3.3 | 0.49952 | 0.49953 | 0.49955 | 0.49957 | 0.49958 | 0.49960 | 0.49961 | 0.49962 | 0.49964 | 0.49965 |
| 3.4 | 0.49966 | 0.49968 | 0.49969 | 0.49970 | 0.49971 | 0.49972 | 0.49973 | 0.49974 | 0.49975 | 0.49976 |
| 3.5 | 0.49977 | 0.49978 | 0.49978 | 0.49979 | 0.49980 | 0.49981 | 0.49981 | 0.49982 | 0.49983 | 0.49983 |
| 3.6 | 0.49984 | 0.49985 | 0.49985 | 0.49986 | 0.49986 | 0.49987 | 0.49987 | 0.49988 | 0.49988 | 0.49989 |
| 3.7 | 0.49989 | 0.49990 | 0.49990 | 0.49990 | 0.49991 | 0.49991 | 0.49992 | 0.49992 | 0.49992 | 0.49992 |
| 3.8 | 0.49993 | 0.49993 | 0.49993 | 0.49994 | 0.49994 | 0.49994 | 0.49994 | 0.49995 | 0.49995 | 0.49995 |
| 3.9 | 0.49995 | 0.49995 | 0.49996 | 0.49996 | 0.49996 | 0.49996 | 0.49996 | 0.49996 | 0.49997 | 0.49997 |
| 4.0 | 0.49997 | 0.49997 | 0.49997 | 0.49997 | 0.49997 | 0.49997 | 0.49998 | 0.49998 | 0.49998 | 0.49998 |

# Table 2: Standard Normal Cumulative Probability Table

Cumulative probabilities for Z - values are given in the following table:

| Z | .00 | .01 | .02 | .03 | .04 | .05 | .06 | .07 | .08 | .09 |
|---|---|---|---|---|---|---|---|---|---|---|
| 0.0 | .5000 | .5040 | .5080 | .5120 | .5160 | .5199 | .5239 | .5279 | .5319 | .5359 |
| 0.1 | .5398 | .5438 | .5478 | .5517 | .5557 | .5596 | .5636 | .5675 | .5714 | .5753 |
| 0.2 | .5793 | .5832 | .5871 | .5910 | .5948 | .5987 | .6026 | .6064 | .6103 | .6141 |
| 0.3 | .6179 | .6217 | .6255 | .6293 | .6331 | .6368 | .6406 | .6443 | .6480 | .6517 |
| 0.4 | .6554 | .6591 | .6628 | .6664 | .6700 | .6736 | .6772 | .6808 | .6844 | .6879 |
| 0.5 | .6915 | .6950 | .6985 | .7019 | .7054 | .7088 | .7123 | .7157 | .7190 | .7224 |
| 0.6 | .7257 | .7291 | .7324 | .7357 | .7389 | .7422 | .7454 | .7486 | .7517 | .7549 |
| 0.7 | .7580 | .7611 | .7642 | .7673 | .7704 | .7734 | .7764 | .7794 | .7823 | .7852 |
| 0.8 | .7881 | .7910 | .7939 | .7967 | .7995 | .8023 | .8051 | .8078 | .8106 | .8133 |
| 0.9 | .8159 | .8186 | .8212 | .8238 | .8264 | .8289 | .8315 | .8340 | .8365 | .8389 |
| 1.0 | .8413 | .8438 | .8461 | .8485 | .8508 | .8531 | .8554 | .8577 | .8599 | .8621 |
| 1.1 | .8643 | .8665 | .8686 | .8708 | .8729 | .8749 | .8770 | .8790 | .8810 | .8830 |
| 1.2 | .8849 | .8869 | .8888 | .8907 | .8925 | .8944 | .8962 | .8980 | .8997 | .9015 |
| 1.3 | .9032 | .9049 | .9066 | .9082 | .9099 | .9115 | .9131 | .9147 | .9162 | .9177 |
| 1.4 | .9192 | .9207 | .9222 | .9236 | .9251 | .9265 | .9279 | .9292 | .9306 | .9319 |
| 1.5 | .9332 | .9345 | .9357 | .9370 | .9382 | .9394 | .9406 | .9418 | .9429 | .9441 |
| 1.6 | .9452 | .9463 | .9474 | .9484 | .9495 | .9505 | .9515 | .9525 | .9535 | .9545 |
| 1.7 | .9554 | .9564 | .9573 | .9582 | .9591 | .9599 | .9608 | .9616 | .9625 | .9633 |
| 1.8 | .9641 | .9649 | .9656 | .9664 | .9671 | .9678 | .9686 | .9693 | .9699 | .9706 |
| 1.9 | .9713 | .9719 | .9726 | .9732 | .9738 | .9744 | .9750 | .9756 | .9761 | .9767 |
| 2.0 | .9772 | .9778 | .9783 | .9788 | .9793 | .9798 | .9803 | .9808 | .9812 | .9817 |
| 2.1 | .9821 | .9826 | .9830 | .9834 | .9838 | .9842 | .9846 | .9850 | .9854 | .9857 |
| 2.2 | .9861 | .9864 | .9868 | .9871 | .9875 | .9878 | .9881 | .9884 | .9887 | .9890 |
| 2.3 | .9893 | .9896 | .9898 | .9901 | .9904 | .9906 | .9909 | .9911 | .9913 | .9916 |
| 2.4 | .9918 | .9920 | .9922 | .9925 | .9927 | .9929 | .9931 | .9932 | .9934 | .9936 |
| 2.5 | .9938 | .9940 | .9941 | .9943 | .9945 | .9946 | .9948 | .9949 | .9951 | .9952 |
| 2.6 | .9953 | .9955 | .9956 | .9957 | .9959 | .9960 | .9961 | .9962 | .9963 | .9964 |
| 2.7 | .9965 | .9966 | .9967 | .9968 | .9969 | .9970 | .9971 | .9972 | .9973 | .9974 |
| 2.8 | .9974 | .9975 | .9976 | .9977 | .9977 | .9978 | .9979 | .9979 | .9980 | .9981 |
| 2.9 | .9981 | .9982 | .9982 | .9983 | .9984 | .9984 | .9985 | .9985 | .9986 | .9986 |
| 3.0 | .9987 | .9987 | .9987 | .9988 | .9988 | .9989 | .9989 | .9989 | .9990 | .9990 |

# Table3: *t*-table

| cum. prob | $t_{.50}$ | $t_{.75}$ | $t_{.80}$ | $t_{.85}$ | $t_{.90}$ | $t_{.95}$ | $t_{.975}$ | $t_{.99}$ | $t_{.995}$ | $t_{.999}$ | $t_{.9995}$ |
|---|---|---|---|---|---|---|---|---|---|---|---|
| one-tail | 0.50 | 0.25 | 0.20 | 0.15 | 0.10 | 0.05 | 0.025 | 0.01 | 0.005 | 0.001 | 0.0005 |
| two-tails | 1.00 | 0.50 | 0.40 | 0.30 | 0.20 | 0.10 | 0.05 | 0.02 | 0.01 | 0.002 | 0.001 |
| df | | | | | | | | | | | |
| 1 | 0.000 | 1.000 | 1.376 | 1.963 | 3.078 | 6.314 | 12.71 | 31.82 | 63.66 | 318.31 | 636.62 |
| 2 | 0.000 | 0.816 | 1.061 | 1.386 | 1.886 | 2.920 | 4.303 | 6.965 | 9.925 | 22.327 | 31.599 |
| 3 | 0.000 | 0.765 | 0.978 | 1.250 | 1.638 | 2.353 | 3.182 | 4.541 | 5.841 | 10.215 | 12.924 |
| 4 | 0.000 | 0.741 | 0.941 | 1.190 | 1.533 | 2.132 | 2.776 | 3.747 | 4.604 | 7.173 | 8.610 |
| 5 | 0.000 | 0.727 | 0.920 | 1.156 | 1.476 | 2.015 | 2.571 | 3.365 | 4.032 | 5.893 | 6.869 |
| 6 | 0.000 | 0.718 | 0.906 | 1.134 | 1.440 | 1.943 | 2.447 | 3.143 | 3.707 | 5.208 | 5.959 |
| 7 | 0.000 | 0.711 | 0.896 | 1.119 | 1.415 | 1.895 | 2.365 | 2.998 | 3.499 | 4.785 | 5.408 |
| 8 | 0.000 | 0.706 | 0.889 | 1.108 | 1.397 | 1.860 | 2.306 | 2.896 | 3.355 | 4.501 | 5.041 |
| 9 | 0.000 | 0.703 | 0.883 | 1.100 | 1.383 | 1.833 | 2.262 | 2.821 | 3.250 | 4.297 | 4.781 |
| 10 | 0.000 | 0.700 | 0.879 | 1.093 | 1.372 | 1.812 | 2.228 | 2.764 | 3.169 | 4.144 | 4.587 |
| 11 | 0.000 | 0.697 | 0.876 | 1.088 | 1.363 | 1.796 | 2.201 | 2.718 | 3.106 | 4.025 | 4.437 |
| 12 | 0.000 | 0.695 | 0.873 | 1.083 | 1.356 | 1.782 | 2.179 | 2.681 | 3.055 | 3.930 | 4.318 |
| 13 | 0.000 | 0.694 | 0.870 | 1.079 | 1.350 | 1.771 | 2.160 | 2.650 | 3.012 | 3.852 | 4.221 |
| 14 | 0.000 | 0.692 | 0.868 | 1.076 | 1.345 | 1.761 | 2.145 | 2.624 | 2.977 | 3.787 | 4.140 |
| 15 | 0.000 | 0.691 | 0.866 | 1.074 | 1.341 | 1.753 | 2.131 | 2.602 | 2.947 | 3.733 | 4.073 |
| 16 | 0.000 | 0.690 | 0.865 | 1.071 | 1.337 | 1.746 | 2.120 | 2.583 | 2.921 | 3.686 | 4.015 |
| 17 | 0.000 | 0.689 | 0.863 | 1.069 | 1.333 | 1.740 | 2.110 | 2.567 | 2.898 | 3.646 | 3.965 |
| 18 | 0.000 | 0.688 | 0.862 | 1.067 | 1.330 | 1.734 | 2.101 | 2.552 | 2.878 | 3.610 | 3.922 |
| 19 | 0.000 | 0.688 | 0.861 | 1.066 | 1.328 | 1.729 | 2.093 | 2.539 | 2.861 | 3.579 | 3.883 |
| 20 | 0.000 | 0.687 | 0.860 | 1.064 | 1.325 | 1.725 | 2.086 | 2.528 | 2.845 | 3.552 | 3.850 |
| 21 | 0.000 | 0.686 | 0.859 | 1.063 | 1.323 | 1.721 | 2.080 | 2.518 | 2.831 | 3.527 | 3.819 |
| 22 | 0.000 | 0.686 | 0.858 | 1.061 | 1.321 | 1.717 | 2.074 | 2.508 | 2.819 | 3.505 | 3.792 |
| 23 | 0.000 | 0.685 | 0.858 | 1.060 | 1.319 | 1.714 | 2.069 | 2.500 | 2.807 | 3.485 | 3.768 |
| 24 | 0.000 | 0.685 | 0.857 | 1.059 | 1.318 | 1.711 | 2.064 | 2.492 | 2.797 | 3.467 | 3.745 |
| 25 | 0.000 | 0.684 | 0.856 | 1.058 | 1.316 | 1.708 | 2.060 | 2.485 | 2.787 | 3.450 | 3.725 |
| 26 | 0.000 | 0.684 | 0.856 | 1.058 | 1.315 | 1.706 | 2.056 | 2.479 | 2.779 | 3.435 | 3.707 |
| 27 | 0.000 | 0.684 | 0.855 | 1.057 | 1.314 | 1.703 | 2.052 | 2.473 | 2.771 | 3.421 | 3.690 |
| 28 | 0.000 | 0.683 | 0.855 | 1.056 | 1.313 | 1.701 | 2.048 | 2.467 | 2.763 | 3.408 | 3.674 |
| 29 | 0.000 | 0.683 | 0.854 | 1.055 | 1.311 | 1.699 | 2.045 | 2.462 | 2.756 | 3.396 | 3.659 |
| 30 | 0.000 | 0.683 | 0.854 | 1.055 | 1.310 | 1.697 | 2.042 | 2.457 | 2.750 | 3.385 | 3.646 |
| 40 | 0.000 | 0.681 | 0.851 | 1.050 | 1.303 | 1.684 | 2.021 | 2.423 | 2.704 | 3.307 | 3.551 |
| 60 | 0.000 | 0.679 | 0.848 | 1.045 | 1.296 | 1.671 | 2.000 | 2.390 | 2.660 | 3.232 | 3.460 |
| 80 | 0.000 | 0.678 | 0.846 | 1.043 | 1.292 | 1.664 | 1.990 | 2.374 | 2.639 | 3.195 | 3.416 |
| 100 | 0.000 | 0.677 | 0.845 | 1.042 | 1.290 | 1.660 | 1.984 | 2.364 | 2.626 | 3.174 | 3.390 |
| 1000 | 0.000 | 0.675 | 0.842 | 1.037 | 1.282 | 1.646 | 1.962 | 2.330 | 2.581 | 3.098 | 3.300 |
| Z | 0.000 | 0.674 | 0.842 | 1.036 | 1.282 | 1.645 | 1.960 | 2.326 | 2.576 | 3.090 | 3.291 |
| | 0% | 50% | 60% | 70% | 80% | 90% | 95% | 98% | 99% | 99.8% | 99.9% |
| | | | | | Confidence Level | | | | | | |

# Table 4: Chi-Square ($\chi^2$) Distribution Table

The shaded area is equal to $\alpha$ for $\chi^2 = \chi^2_\alpha$.

| df | $\chi^2_{.995}$ | $\chi^2_{.990}$ | $\chi^2_{.975}$ | $\chi^2_{.950}$ | $\chi^2_{.900}$ | $\chi^2_{.100}$ | $\chi^2_{.050}$ | $\chi^2_{.025}$ | $\chi^2_{.010}$ | $\chi^2_{.005}$ |
|---|---|---|---|---|---|---|---|---|---|---|
| 1 | 0.000 | 0.000 | 0.001 | 0.004 | 0.016 | 2.706 | 3.841 | 5.024 | 6.635 | 7.879 |
| 2 | 0.010 | 0.020 | 0.051 | 0.103 | 0.211 | 4.605 | 5.991 | 7.378 | 9.210 | 10.597 |
| 3 | 0.072 | 0.115 | 0.216 | 0.352 | 0.584 | 6.251 | 7.815 | 9.348 | 11.345 | 12.838 |
| 4 | 0.207 | 0.297 | 0.484 | 0.711 | 1.064 | 7.779 | 9.488 | 11.143 | 13.277 | 14.860 |
| 5 | 0.412 | 0.554 | 0.831 | 1.145 | 1.610 | 9.236 | 11.070 | 12.833 | 15.086 | 16.750 |
| 6 | 0.676 | 0.872 | 1.237 | 1.635 | 2.204 | 10.645 | 12.592 | 14.449 | 16.812 | 18.548 |
| 7 | 0.989 | 1.239 | 1.690 | 2.167 | 2.833 | 12.017 | 14.067 | 16.013 | 18.475 | 20.278 |
| 8 | 1.344 | 1.646 | 2.180 | 2.733 | 3.490 | 13.362 | 15.507 | 17.535 | 20.090 | 21.955 |
| 9 | 1.735 | 2.088 | 2.700 | 3.325 | 4.168 | 14.684 | 16.919 | 19.023 | 21.666 | 23.589 |
| 10 | 2.156 | 2.558 | 3.247 | 3.940 | 4.865 | 15.987 | 18.307 | 20.483 | 23.209 | 25.188 |
| 11 | 2.603 | 3.053 | 3.816 | 4.575 | 5.578 | 17.275 | 19.675 | 21.920 | 24.725 | 26.757 |
| 12 | 3.074 | 3.571 | 4.404 | 5.226 | 6.304 | 18.549 | 21.026 | 23.337 | 26.217 | 28.300 |
| 13 | 3.565 | 4.107 | 5.009 | 5.892 | 7.042 | 19.812 | 22.362 | 24.736 | 27.688 | 29.819 |
| 14 | 4.075 | 4.660 | 5.629 | 6.571 | 7.790 | 21.064 | 23.685 | 26.119 | 29.141 | 31.319 |
| 15 | 4.601 | 5.229 | 6.262 | 7.261 | 8.547 | 22.307 | 24.996 | 27.488 | 30.578 | 32.801 |
| 16 | 5.142 | 5.812 | 6.908 | 7.962 | 9.312 | 23.542 | 26.296 | 28.845 | 32.000 | 34.267 |
| 17 | 5.697 | 6.408 | 7.564 | 8.672 | 10.085 | 24.769 | 27.587 | 30.191 | 33.409 | 35.718 |
| 18 | 6.265 | 7.015 | 8.231 | 9.390 | 10.865 | 25.989 | 28.869 | 31.526 | 34.805 | 37.156 |
| 19 | 6.844 | 7.633 | 8.907 | 10.117 | 11.651 | 27.204 | 30.144 | 32.852 | 36.191 | 38.582 |
| 20 | 7.434 | 8.260 | 9.591 | 10.851 | 12.443 | 28.412 | 31.410 | 34.170 | 37.566 | 39.997 |
| 21 | 8.034 | 8.897 | 10.283 | 11.591 | 13.240 | 29.615 | 32.671 | 35.479 | 38.932 | 41.401 |
| 22 | 8.643 | 9.542 | 10.982 | 12.338 | 14.041 | 30.813 | 33.924 | 36.781 | 40.289 | 42.796 |
| 23 | 9.260 | 10.196 | 11.689 | 13.091 | 14.848 | 32.007 | 35.172 | 38.076 | 41.638 | 44.181 |
| 24 | 9.886 | 10.856 | 12.401 | 13.848 | 15.659 | 33.196 | 36.415 | 39.364 | 42.980 | 45.559 |
| 25 | 10.520 | 11.524 | 13.120 | 14.611 | 16.473 | 34.382 | 37.652 | 40.646 | 44.314 | 46.928 |
| 26 | 11.160 | 12.198 | 13.844 | 15.379 | 17.292 | 35.563 | 38.885 | 41.923 | 45.642 | 48.290 |
| 27 | 11.808 | 12.879 | 14.573 | 16.151 | 18.114 | 36.741 | 40.113 | 43.195 | 46.963 | 49.645 |
| 28 | 12.461 | 13.565 | 15.308 | 16.928 | 18.939 | 37.916 | 41.337 | 44.461 | 48.278 | 50.993 |
| 29 | 13.121 | 14.256 | 16.047 | 17.708 | 19.768 | 39.087 | 42.557 | 45.722 | 49.588 | 52.336 |
| 30 | 13.787 | 14.953 | 16.791 | 18.493 | 20.599 | 40.256 | 43.773 | 46.979 | 50.892 | 53.672 |
| 40 | 20.707 | 22.164 | 24.433 | 26.509 | 29.051 | 51.805 | 55.758 | 59.342 | 63.691 | 66.766 |
| 50 | 27.991 | 29.707 | 32.357 | 34.764 | 37.689 | 63.167 | 67.505 | 71.420 | 76.154 | 79.490 |
| 60 | 35.534 | 37.485 | 40.482 | 43.188 | 46.459 | 74.397 | 79.082 | 83.298 | 88.379 | 91.952 |
| 70 | 43.275 | 45.442 | 48.758 | 51.739 | 55.329 | 85.527 | 90.531 | 95.023 | 100.425 | 104.215 |
| 80 | 51.172 | 53.540 | 57.153 | 60.391 | 64.278 | 96.578 | 101.879 | 106.629 | 112.329 | 116.321 |
| 90 | 59.196 | 61.754 | 65.647 | 69.126 | 73.291 | 107.565 | 113.145 | 118.136 | 124.116 | 128.299 |
| 100 | 67.328 | 70.065 | 74.222 | 77.929 | 82.358 | 118.498 | 124.342 | 129.561 | 135.807 | 140.169 |

# Table 5: Critical Values of the F Distribution with Alpha Level of 0.05

| $v_2$ \ $v_1$ | 1 | 2 | 3 | 4 | 5 | 6 | 7 | 8 | 9 | 10 | 12 | 15 | 20 | 24 | 30 | 40 | 60 | 120 | ∞ |
|---|---|---|---|---|---|---|---|---|---|---|---|---|---|---|---|---|---|---|---|
| 1 | 161.4 | 199.5 | 215.7 | 224.6 | 230.2 | 234.0 | 236.8 | 238.9 | 240.5 | 241.9 | 243.9 | 245.9 | 248.0 | 249.1 | 250.1 | 251.1 | 252.2 | 253.3 | 254.3 |
| 2 | 18.51 | 19.00 | 19.16 | 19.25 | 19.30 | 19.33 | 19.35 | 19.37 | 19.38 | 19.40 | 19.41 | 19.43 | 19.45 | 19.45 | 19.46 | 19.47 | 19.48 | 19.49 | 19.50 |
| 3 | 10.13 | 9.55 | 9.28 | 9.12 | 9.01 | 8.94 | 8.89 | 8.85 | 8.81 | 8.79 | 8.74 | 8.70 | 8.66 | 8.64 | 8.62 | 8.59 | 8.57 | 8.55 | 8.53 |
| 4 | 7.71 | 6.94 | 6.59 | 6.39 | 6.26 | 6.16 | 6.09 | 6.04 | 6.00 | 5.96 | 5.91 | 5.86 | 5.80 | 5.77 | 5.75 | 5.72 | 5.69 | 5.66 | 5.63 |
| 5 | 6.61 | 5.79 | 5.41 | 5.19 | 5.05 | 4.95 | 4.88 | 4.82 | 4.77 | 4.74 | 4.68 | 4.62 | 4.56 | 4.53 | 4.50 | 4.46 | 4.43 | 4.40 | 4.36 |
| 6 | 5.99 | 5.14 | 4.76 | 4.53 | 4.39 | 4.28 | 4.21 | 4.15 | 4.10 | 4.06 | 4.00 | 3.94 | 3.87 | 3.84 | 3.81 | 3.77 | 3.74 | 3.70 | 3.67 |
| 7 | 5.59 | 4.74 | 4.35 | 4.12 | 3.97 | 3.87 | 3.79 | 3.73 | 3.68 | 3.64 | 3.57 | 3.51 | 3.44 | 3.41 | 3.38 | 3.34 | 3.30 | 3.27 | 3.23 |
| 8 | 5.32 | 4.46 | 4.07 | 3.84 | 3.69 | 3.58 | 3.50 | 3.44 | 3.39 | 3.35 | 3.28 | 3.22 | 3.15 | 3.12 | 3.08 | 3.04 | 3.01 | 2.97 | 2.93 |
| 9 | 5.12 | 4.26 | 3.86 | 3.63 | 3.48 | 3.37 | 3.29 | 3.23 | 3.18 | 3.14 | 3.07 | 3.01 | 2.94 | 2.90 | 2.86 | 2.83 | 2.79 | 2.75 | 2.71 |
| 10 | 4.96 | 4.10 | 3.71 | 3.48 | 3.33 | 3.22 | 3.14 | 3.07 | 3.02 | 2.98 | 2.91 | 2.85 | 2.77 | 2.74 | 2.70 | 2.66 | 2.62 | 2.58 | 2.54 |
| 11 | 4.84 | 3.98 | 3.59 | 3.36 | 3.20 | 3.09 | 3.01 | 2.95 | 2.90 | 2.85 | 2.79 | 2.72 | 2.65 | 2.61 | 2.57 | 2.53 | 2.49 | 2.45 | 2.40 |
| 12 | 4.75 | 3.89 | 3.49 | 3.26 | 3.11 | 3.00 | 2.91 | 2.85 | 2.80 | 2.75 | 2.69 | 2.62 | 2.54 | 2.51 | 2.47 | 2.43 | 2.38 | 2.34 | 2.30 |
| 13 | 4.67 | 3.81 | 3.41 | 3.18 | 3.03 | 2.92 | 2.83 | 2.77 | 2.71 | 2.67 | 2.60 | 2.53 | 2.46 | 2.42 | 2.38 | 2.34 | 2.30 | 2.25 | 2.21 |
| 14 | 4.60 | 3.74 | 3.34 | 3.11 | 2.96 | 2.85 | 2.76 | 2.70 | 2.65 | 2.60 | 2.53 | 2.46 | 2.39 | 2.35 | 2.31 | 2.27 | 2.22 | 2.18 | 2.13 |
| 15 | 4.54 | 3.68 | 3.29 | 3.06 | 2.90 | 2.79 | 2.71 | 2.64 | 2.59 | 2.54 | 2.48 | 2.40 | 2.33 | 2.29 | 2.25 | 2.20 | 2.16 | 2.11 | 2.07 |
| 16 | 4.49 | 3.63 | 3.24 | 3.01 | 2.85 | 2.74 | 2.66 | 2.59 | 2.54 | 2.49 | 2.42 | 2.35 | 2.28 | 2.24 | 2.19 | 2.15 | 2.11 | 2.06 | 2.01 |
| 17 | 4.45 | 3.59 | 3.20 | 2.96 | 2.81 | 2.70 | 2.61 | 2.55 | 2.49 | 2.45 | 2.38 | 2.31 | 2.23 | 2.19 | 2.15 | 2.10 | 2.06 | 2.01 | 1.96 |
| 18 | 4.41 | 3.55 | 3.16 | 2.93 | 2.77 | 2.66 | 2.58 | 2.51 | 2.46 | 2.41 | 2.34 | 2.27 | 2.19 | 2.15 | 2.11 | 2.06 | 2.02 | 1.97 | 1.92 |
| 19 | 4.38 | 3.52 | 3.13 | 2.90 | 2.74 | 2.63 | 2.54 | 2.48 | 2.42 | 2.38 | 2.31 | 2.23 | 2.16 | 2.11 | 2.07 | 2.03 | 1.98 | 1.93 | 1.88 |
| 20 | 4.35 | 3.49 | 3.10 | 2.87 | 2.71 | 2.60 | 2.51 | 2.45 | 2.39 | 2.35 | 2.28 | 2.20 | 2.12 | 2.08 | 2.04 | 1.99 | 1.95 | 1.90 | 1.84 |
| 21 | 4.32 | 3.47 | 3.07 | 2.84 | 2.68 | 2.57 | 2.49 | 2.42 | 2.37 | 2.32 | 2.25 | 2.18 | 2.10 | 2.05 | 2.01 | 1.96 | 1.92 | 1.87 | 1.81 |
| 22 | 4.30 | 3.44 | 3.05 | 2.82 | 2.66 | 2.55 | 2.46 | 2.40 | 2.34 | 2.30 | 2.23 | 2.15 | 2.07 | 2.03 | 1.98 | 1.94 | 1.89 | 1.84 | 1.78 |
| 23 | 4.28 | 3.42 | 3.03 | 2.80 | 2.64 | 2.53 | 2.44 | 2.37 | 2.32 | 2.27 | 2.20 | 2.13 | 2.05 | 2.01 | 1.96 | 1.91 | 1.86 | 1.81 | 1.76 |
| 24 | 4.26 | 3.40 | 3.01 | 2.78 | 2.62 | 2.51 | 2.42 | 2.36 | 2.30 | 2.25 | 2.18 | 2.11 | 2.03 | 1.98 | 1.94 | 1.89 | 1.84 | 1.79 | 1.73 |
| 25 | 4.24 | 3.39 | 2.99 | 2.76 | 2.60 | 2.49 | 2.40 | 2.34 | 2.28 | 2.24 | 2.16 | 2.09 | 2.01 | 1.96 | 1.92 | 1.87 | 1.82 | 1.77 | 1.71 |
| 26 | 4.23 | 3.37 | 2.98 | 2.74 | 2.59 | 2.47 | 2.39 | 2.32 | 2.27 | 2.22 | 2.15 | 2.07 | 1.99 | 1.95 | 1.90 | 1.85 | 1.80 | 1.75 | 1.69 |
| 27 | 4.21 | 3.35 | 2.96 | 2.73 | 2.57 | 2.46 | 2.37 | 2.31 | 2.25 | 2.20 | 2.13 | 2.06 | 1.97 | 1.93 | 1.88 | 1.84 | 1.79 | 1.73 | 1.67 |
| 28 | 4.20 | 3.34 | 2.95 | 2.71 | 2.56 | 2.45 | 2.36 | 2.29 | 2.24 | 2.19 | 2.12 | 2.04 | 1.96 | 1.91 | 1.87 | 1.82 | 1.77 | 1.71 | 1.65 |
| 29 | 4.18 | 3.33 | 2.93 | 2.70 | 2.55 | 2.43 | 2.35 | 2.28 | 2.22 | 2.18 | 2.10 | 2.03 | 1.94 | 1.90 | 1.85 | 1.81 | 1.75 | 1.70 | 1.64 |
| 30 | 4.17 | 3.32 | 2.92 | 2.69 | 2.53 | 2.42 | 2.33 | 2.27 | 2.21 | 2.16 | 2.09 | 2.01 | 1.93 | 1.89 | 1.84 | 1.79 | 1.74 | 1.68 | 1.62 |
| 40 | 4.08 | 3.23 | 2.84 | 2.61 | 2.45 | 2.34 | 2.25 | 2.18 | 2.12 | 2.08 | 2.00 | 1.92 | 1.84 | 1.79 | 1.74 | 1.69 | 1.64 | 1.58 | 1.51 |
| 60 | 4.00 | 3.15 | 2.76 | 2.53 | 2.37 | 2.25 | 2.17 | 2.10 | 2.04 | 1.99 | 1.92 | 1.84 | 1.75 | 1.70 | 1.65 | 1.59 | 1.53 | 1.47 | 1.39 |
| 120 | 3.92 | 3.07 | 2.68 | 2.45 | 2.29 | 2.17 | 2.09 | 2.02 | 1.96 | 1.91 | 1.83 | 1.75 | 1.66 | 1.61 | 1.55 | 1.50 | 1.43 | 1.35 | 1.25 |
| ∞ | 3.84 | 3.00 | 2.60 | 2.37 | 2.21 | 2.10 | 2.01 | 1.94 | 1.88 | 1.83 | 1.75 | 1.67 | 1.57 | 1.52 | 1.46 | 1.39 | 1.32 | 1.22 | 1.00 |

www.ingramcontent.com/pod-product-compliance
Lightning Source LLC
Chambersburg PA
CBHW062108220526
45471CB00010B/3647